Crickets and Katydids, Concerts and Solos

Crickets and Katydids, Concerts and Solos

Vincent G. Dethier

Foreword by A. R. Ammons

Harvard University Press
Cambridge, Massachusetts
London, England   1992

TEXT ILLUSTRATIONS BY ABIGAIL RORER.

"How dreary—to be—somebody!" is reprinted by permission of the publishers and the
Trustees of Amherst College from *The Poems of Emily Dickinson*, Thomas H. Johnson, ed.,
Cambridge, Mass.: The Belknap Press of Harvard University Press, copyright 1951, ©
1955, 1979, 1983 by the President and Fellows of Harvard College.

Poem by Basho is from *More Cricket Songs*, Japanese haiku translated by Harry Behn.
Copyright © 1971 by Harry Behn. Reprinted by permission of Marian Reiner.

This book is printed on acid-free paper, and its binding materials
have been chosen for strength and durability.

Library of Congress Cataloging-in-Publication Data

Dethier, V. G. (Vincent Gaston), 1915–
    Crickets and katydids, concerts and solos / Vincent G. Dethier.
        p.   cm.
    Includes bibliographical references.
    ISBN 0–674–17577–8 (alk. paper)
    1. Crickets—Northeastern States—Behavior. 2. Crickets—
Northeastern States—Vocalization. 3. Katydids—Northeastern
States—Behavior. 4. Katydids—Northeastern States—Vocalization.
5. Grasshoppers—Northeastern States—Behavior. 6. Grasshoppers—
Northeastern States—Vocalization. 7. Insect sounds—Northeastern
States. I. Title.
QL508.G8D48 1992
595.7'260459—dc20
92–3718
        CIP

Foreword

by A. R. Ammons

*T*wo things are dead giveaways in nature: one is moving, making a motion, and the other is making a sound. Because these two are so risky, wild nature is mostly still and quiet, especially if we consider how many creatures are there and how much "business" has to be transacted—just about all the business that we would consider our business: obtaining food and housing (nesting), breeding, and, for some, brooding or rearing young. So many actions in our lives, as in the lives of insects and other animals, take their value from and seem to be governed by risk, nothing easy, free, or meaningful otherwise.

Vincent Dethier shows us how to listen for sound in fields, edges, and woods and to become aware of the movements that accompany sound. We pick up on the behavior of Orthoptera and acquire a sense of the "goals" or "ends"—the narratives—that underlie the sounds. A science of sensation is just as exacting and requires just as much attention and insight as do the sciences of mass and number. It is, after all, the information from our senses that we live in, and while unaided they may limit our knowledge of microbes and nebulae, they on the other hand offer us the fullness of existence.

A very special reward of Mr. Dethier's book is the sound *he* makes, in rich and revealing language. We learn from his sounds what kind of person, capable of this kind of interest and care, is attending to our minds. His own sound becomes part of the community of sound common to most, if not nearly all, life, so we are doubly trained to hear, and we become doubly committed to understanding and caring for all forms of life.

Contents

Illustrations

Preface

This book is an invitation to share a listening pleasure that will beguile away the lazy days of summer, the bittersweet golden days of Indian Summer, and the nostalgic days of fall. It is an introduction to some of the more common singing crickets, locusts, and grasshoppers of the northeastern United States. In describing the behavior of these insects, it reveals tangentially something of the behavior of those who study them. Which of the two behaviors is the more interesting I leave to you to determine.

These small essays make no pretense of being comprehensive either in terms of the number of species treated or in the depth of the biological information about each species. This book is not a field guide, nor is it an entomological treatise. For the reader who wishes to delve more deeply into the classification, structure, or habits of particular crickets or grasshoppers, a selected list for further reading is appended. These works include detailed keys for identification, treatments of species living beyond the bounds of New England, and technical analyses of songs. For comprehensive information, the books of Blatchley and Morse are highly recommended. Although difficult to obtain, they are worth the effort of a search because they are richer in biological detail than most of the more up-to-date works. The latter are also included in the list of references.

A general treatment of sound production, recording techniques, and sound analyses is given in G. W. Pierce's pioneering book *Songs of Insects*. In Pierce's own words, his observations on the sounds emitted by insects "are incidental to the wider interest of the study, which is acoustics, particularly in the supersonic range of frequencies." Pierce's study extended over a period of twelve years; I was privileged to work with him for three. My involvement was merely that of entomological factotum.

The vignettes presented here condense the experience of three years into one. They constitute an idyll centered on the studies in Professor Pierce's laboratory. The collection is dedicated to his memory.

Crickets and Katydids, Concerts and Solos

Prelude

The Time hath laid his mantle by
Of wind and rain and icy chill,
And dons a rich embroidery
Of sun-light pour'd on lake and hill.
No beast or bird in earth or sky
Whose voice doth not with gladness thrill,
For Time hath laid his mantle by
Of wind and rain and icy chill.
Charles d'Orléans

Spring belongs to the frogs and toads, the spring peepers, the knee-deeps, however you choose to name them. Relatives all, they fill the night with shrilling songs of love. What passion is this that moves them to crouch in water crystalline with the remembrance of winter! The solitude of the swamps pulsates with their mating calls while the rest of nature is waking but slowly from winter sleep. Since the ancient change from the primeval inanimate chemical world to life began in water, it is fitting that the reawakening each year should begin in the swamps and swales and that the jubilee should be celebrated there.

The fields are sere and matted, the grass not yet greening, but the willows are yellowing and the osiers reddening. The wild blackberry canes are accented in mauve. Denizens of the fields sleep undisturbed in detritus and earthen chambers. In lawns and gardens a groundskeeper plying his rake combs out here and there a woolly-bear caterpillar, but in its torpor it remains coiled head to tail. Truly, spring belongs to the frogs and toads. Except for them, nature has not found her voice.

Almost imperceptibly the days lengthen. A suspicion of warmth tempers the air. In the woods the downy and hairy woodpeckers drum messages from favorite seasoned dead limbs. The redwings take possession of the swamps. From swaying perches on the skeletal stems of last year's cattails, they "cheerokee" to each other. In the bordering hedges the first song sparrows practice their scales. Robins have arrived but have yet to sing. By the end of May most of New England's feathered summer residents have arrived, have found their voices. They warble and trill from the treetops, the fence posts, and the deepness of the woods. They occupy nature's front stage. The last chill is dispelled by the sun. The late spring and early summer days belong to the birds.

By early June the spring peepers have abandoned their chorusing for solitary living. The songs of the birds are muted and no longer have the same note of spontaneity and urgency as at the start of nesting season. There comes a pause in nature's score. Oh, the birds still sing, but the songs sound a bit jaded, tempered with post-nuptial staidness; there is even a hint of ennui. Compared to the sounds of spring there is a musical diminuendo, even a long musical rest presaging a change in key, a change in tempi, a shift in mood and theme. Only a caucus of crows still disputes in the pine woods.

Few people are acquainted with the singers which will replace the birds. They may have read of the cricket on the hearth, or heard the arguments of the katydids, or noted in passing the "hot bugs," the cicadas, but unless they have tuned their ears to the level of the grasses or cocked them to the thickets, the richness of the insect repertoire is hidden from their consciousness. We who are preeminently creatures of vision instinctively appreciate most the world of light; we must learn to savor the full range of nature's world of sound.

Sad to say, full appreciation of summer's music is reserved for the young, because the pitch of so many of the songs is too high to be perceived by middle-aged and elderly ears. This failing is just one facet of the dulling of the senses with age, perhaps a divinely calculated diminution so that we may withdraw from the world gradually.

For those young enough to hear, the discovery of the singers of midsummer and autumn carries with it the expectancies of exotic explorations and the intrigues of detective stories. The reward for a successful encounter is the pleasure of a novel experience, aesthetic gratification, and the satisfaction of having stalked and outwitted an unknown singer that treasures his anonymity. The singing insects are wary, alert, clever in the art of concealment, and as deceptive as the most skillful ventriloquists.

Each person who is at one with nature has his own personally characteristic introduction to the singing crickets, locusts, and katydids, known collectively as Orthoptera, "the straight-winged ones." If you are a neophyte who knows nothing of these singers (or, more properly, fiddlers, since their songs are not produced vocally), and you have never stalked them, the experience that awaits you may be savored vicariously through a brief chronicle of one person's introduction.

In much the same way as I began, you will commence your exploration, build a repertory of known songs, and hear nature's world of sound expand as each singer's identity and song are stored in your memory and each familiar solo suddenly rings loud and informative against the background of your increasing knowledge. And then, as summer advances, any new song will immediately alert you, and the game of hide-and-seek will begin anew.

My introduction to this world of trills, tinklings, buzzes, shuffles, crackles, and clicks occurred purely by chance—a mixture of fortuitous circumstance, an older man's wonderment, my young ears, and one college course in entomology. Opportunity arrived in the form of a telephone call. At that time, in the early thirties, the Great Depression still cast its shadow. I was finishing college and preparing to enter graduate school. The call could not have come at a more opportune moment: it was an offer of summer employment. The person interested in my services was Professor George W. Pierce, an emeritus professor of physics at Harvard.

The philosopher may well wonder whether or not there is

any sound if a tree crashes in the forest and no one hears it fall. Professor Pierce had conceived a far more fruitful curiosity regarding sound. He had been wondering whether or not there were in nature sounds of such high pitch that human beings were unaware of their existence. Was nature in fact an incredibly noisy place?

Pierce's research specialty was the production and detection of high-frequency sound. During World War I, in connection with antisubmarine defense and underwater communication, he had conducted for the United States Navy research relating to the production, transmission, and detection of sound signals in air and in water. Then and afterward he had developed apparatus capable of detecting high-frequency sound and transforming it so that it could be heard, recorded, and analyzed.

With these devices he began to listen to singing birds. Birds proved to be disappointing subjects because the more interesting characteristics of their songs were audible to the human ear. There were few surprises. Then his interest turned to the singing Orthoptera. In 1936 he built a laboratory at his summer home in Franklin, New Hampshire, on part of an old abandoned farm purchased in 1910 by Harvard physiologist Walter B. Cannon. The Cannons lived in the farmhouse, which was located on a high ridge overlooking the Pemigewasset Valley, with Mount Kearsarge on the westerly horizon and Cardigan mountain on the northern. In the summer of 1938, at Professor Pierce's invitation, I packed my belongings in the rumble seat of a Model A Ford and left Boston for Franklin, where I was to collect, identify, and care for as many kinds of crickets, locusts, and katydids as could be found in that part of New Hampshire. Pierce had also hired a photographer, Paul Donaldson.

Franklin lies at the point where the stream draining Lake Webster meets the Pemigewasset River; it is here that the Pemigewasset becomes the Merrimack, just south and west of Lake Winnipesaukee. The Pierces had built their summer home on the site of a former chalet constructed by the Cannons. A small frame laboratory, raw white pine within and white painted clapboard without, was situated a few yards downhill of the main house. It sheltered beneath a clump of young rock maples, hard by the usual New England wall of field stones. An open porch on the downhill side looked out over hayfields and old pastures, which slid away to a dark, cool vale. There a small brook burbled over ledges and through a sphagnum bog. On the far side of the laboratory, out of sight of the porch, the land rose to higher pastures, dry and sunny, and spotted with ledges, clumps of bracken, extensive patches of light green hay fern, and thickets of sweet fern (*Comptonia*) and bayberry. Mount Kearsarge rose on the horizon beyond.

These details are significant because even though I did not realize it at the time, here were all the varied habitats, save salt marshes, congenial to the immense variety of Orthoptera of that latitude. These insects, all built on the same general plan, with tremendous sinewy jumping hind legs and appetites for all sorts of grasses and sedges, were, nonetheless, each adapted to thrum in a special habitat, from the cool, damp sphagnum through all ranges of light and humidity to scorching exposed granite ledges. There was an orthopteran for every habitat, and each habitat was populated by its own well-adjusted citizens. In time, Donaldson and I came to know just where to search for each particular species, under what weather conditions, and at what time of day or night.

Though I had not suspected it earlier, I learned that collecting requires immense experience and skill. It is a pursuit that cannot be learned from books. One develops a search image—that is, a generalized mental picture of body shape and color; one learns the preferred spots in the habitat, a leaf, a stem, the top or midpoint of a plant. Even more crucial, one learns the characteristic escape tactics of each species. One species freezes into the background, relying on camouflage and immobility to save it. Another takes flight. Still another drops to the tangle of detritus.

The field laboratory at Franklin, New Hampshire

Some sidle around to the far side of a stem, others squeeze themselves into crevices. The weak flyers soon come to ground; the strong ones are off and away. The pursuer learns to swing the net from ahead of the quarry. He learns which clump of bracken to sweep and whether to sweep grasses high or low.

The interior of the laboratory was simplicity itself, an encouragement to gifted amateurs that shows how much can be accomplished with bare essentials. It was frugally designed to accommodate the catch and provide equipment for detailed studies of form and behavior. There was a small closet that served as a photographic darkroom. The rest of the building consisted of a workbench extending along one wall, shelves on the remaining walls, and several cane-bottomed chairs. Two microscopes, several lamps, apparatus for recording insect songs, a large number of cages built by a local carpenter, and one portable electric heater completed the inventory. During the chilly days of autumn, the heater gave an illusion of warmth. Its real purpose, however, was to warm the singers and bring them up to tempo when songs were desired. The only decorations were several striking oil portraits of Conehead Meadow Grasshoppers, painted by Pierce himself.

My duties were straightforward. I was to collect living specimens of all the singing Orthoptera endemic to that part of New Hampshire, identify them, and keep them sufficiently well fed and content in their cages so that they would sing with gusto. Until that summer I had never studied crickets or grasshoppers. My bibles and my salvation were the classic *The Orthoptera of Northeastern America, with Special Reference to the Faunas of Indiana and Florida*, by W. S. Blatchley, and A. P. Morse's *Manual of the Orthoptera of New England*. The keys for description, together with the notes on geographic range and habitat, proved invaluable. Eventually I came to master, almost, the idiosyncratic distinctions that taxonomists made among species. I learned that genitalia were the Rosetta stones of classification—a quirk that always intrigued me because it seemed odd that such trivialities as short spine versus long spine on part of the external genital apparatus should be so diagnostic while the more obvious characters seemed so unimportant and so variable. What did "long" and "short" mean anyway? In time I learned to recognize species without quite knowing what the diagnostic characters were, just as one recognizes members of one's family not by individual characteristics but by the whole.

There were further difficulties. For instance, the identities, or at least the names, of insects kept changing as specialists studied and restudied species. As you will discover in the following pages, the familiar scientific names of many crickets and locusts have been replaced. As a consequence, it is sometimes difficult for the amateur to relate the contemporary names to species of the older literature.

Paul Donaldson was responsible for all matters photographic—taking still pictures, and colored motion pictures at various speeds, developing and printing pictures and sound records, and sharing with me the duties of factotum. Apart from these assignments, we were free to do as we pleased. We reported each morning to the laboratory, and unless there was some special project afoot, such as recording songs or assisting Professor Pierce in analyzing records, we spent most of our days in the fields, where we roamed and looked and listened. A more idyllic occupation would be hard to imagine. Under what other circumstances could a person lie on his back staring at the marching cumulus and be working, while to all appearances he was loafing? These are pleasures that can be shared vicariously.

The Field Crickets

Trapped in a helmet
hung in a shrine, a cricket
Chirps his last command.
Basho, tr. Harry Behn

When we arrived in Franklin early in June to set things in order, the fields were long-haired and lush. Late spring flowers had yielded their places to early summer flowers. Dandelions in golden array dominated the grasslands. Daisies were not yet in bloom, but wild strawberries had already added a glow where the grass was short. In closely cropped pastures buttercups clumped in conspicuous isolation, carefully avoided by the cows, which had some instinctive sense of their poisonous nature. In shady moist hedgerows, the purple pyrola was faded and the arbutus already past. The Jack-in-the-Pulpits in the woods were now high and sturdy. Shoots of corn in cultivated fields had pushed three inches above the black soil.

On this particular morning, tentative, exquisitely gentle breezes tested the suppleness of seed-laden grasses. The sighs of wind were too hesitant to sway branches of shrubs and trees, but now and again a single maple leaf would pivot on its stem and wave like a signaling hand. Above the greenery the season's first aphids drifted by, tiny motes able to sustain themselves aloft with a minimum expenditure of energy. From a clump of poplars an incongruous snowstorm of catkin fluff bore the destiny of the next generation wherever the vagaries of the currents decreed.

As the morning wore on, I interrupted my communion with nature with reluctant spasms of honest hard labor. By noontime the laboratory was in operational readiness, but there was as yet no scientific business requiring attention. Since my arrival I had not heard a single insect, but there would be time enough for that. Comforted by that thought, I walked a short distance down the slope below the laboratory to eat a frugal lunch. I found a convenient stump, stretched my legs in the grass, chewed my sandwich, and contemplated the terrain of my future labors.

It was a day for dreaming. The breeze had died. Aloft, towering castles of cumulus seemed to change neither their shape nor their station. Looking over the fields was like viewing a Dutch landscape painting, where sky always dominates terrain. It was a supremely serene day, and the most quiet hour of that day. Suddenly a cricket began to chirp!

Soon he was joined by another, and then by others. I realized that summer had indeed arrived—not the equinoctial summer but the summer of animated nature. All at once the day belonged to the crickets.

The synchrony of season and advent of maturity is one of the universal wonders of nature. One day the crickets are mute nymphs; then, as if by some agreed-upon signal cued by the season, they all molt, become adults, and celebrate the occasion in song. We know, of course, that the synchrony is hormonally controlled and that the hormonal glands are obedient to temperature and day/night cycles. We know that the score of the music is inscribed in the developing nervous system and awaits only the construction of an instrument for its expression. We know, too, that each score is unique to its genetic composer and its performance possible only by a single species. None of this diminishes the wonder or the mystery.

The debut of the crickets was timed to the fullness of the fields and the flowering of the waste places. Frogs were croaking in the ponds, but the crickets were beginning to upstage them. From this moment on, the crickets would sing throughout the afternoon and throughout the night. They would cease only when the robins began calling to awaken the sun. By mid-morning, when the birds fell silent, the crickets would resume their concert-izing. Only rain, excessive heat, or chill would silence them.

These, the first singing orthopterans of the year, are the larg-est crickets in the northeast. The words that best describe them are "stout," "round," "robust." Their attire, like the formal attire of male concert soloists, is black, relieved only by blending shades of brown on their wing covers.

Of all the Orthoptera with which I became acquainted dur-ing the three summers of research in Franklin, the field crickets remain in my memory as the most attractive. Perhaps the appeal lies in their round faces; by contrast, the singing grasshoppers and locusts have narrow, lean, hungry-looking faces. Field crickets also have versatile, individualistic behaviors.

Many people view these large black crickets with the same distaste they accord cockroaches. The most charitable explana-tion of this attitude is that most people encounter field crickets, especially their immigrant cousins from Europe, the domestic field crickets, out of their natural habitat—that is to say, in houses. There they are perceived as "big, black, disgusting bugs," unwel-come interlopers, vermin. At the other end of the psychological spectrum, there are different feelings. The Chinese see crickets as valiant gladiators in the arena of professional cricket fighting. The Japanese keep them in elaborate small cages as cheery pets, or as "watchdogs" that alert the householder to the presence of burglars by abruptly ceasing their singing. In the Western world the "cricket on the hearth" is celebrated as an omen of good luck, a token of conviviality and companionship.

In his own world this cricket is a hale, hearty, independent fellow. Each male is lord of his own modest territory: a burrow under a clump of grass, a chink in a wall, a crack in a log, a crevice in the soil, a lodge among the roots of a shrub, a shelter in the shade of a rosette of dandelion leaves. However humble his dom-icile, it is often his castle for life. He chirps there almost contin-

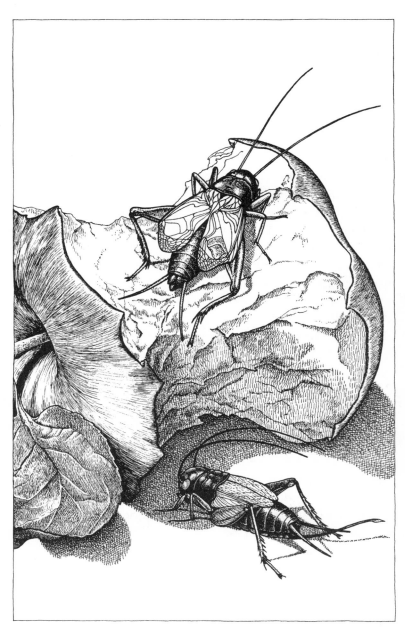

Spring field crickets

uously, until one would think that he would collapse from exhaustion or his instrument would be worn to silence.

The instrument of this indefatigable musician is his interlocking front wings, the stiff parchment-like covers for his membranous hind wings. As the bow of a violin is drawn across the strings and sets them vibrating and as the body of the violin is set resonating by transmission of the vibrations through the bridge, so the cricket draws a scraper across a file of small teeth and sets the wing covers to resonating. The wing covers are raised in song at an angle of about forty-five degrees and brought together periodically. With this closing motion, a ridge on the upper wing cover scrapes across a file on the lower wing cover and generates a high-pitched sound lasting about one hundredth of a second. Each chirp consists of a group of three or more of these sound pulses. They are executed so rapidly that the human ear cannot resolve the pattern. We hear only the chirp. The repetition of the chirp makes up the song. The song I heard that day in early June was the calling song.

Sometimes the troubador becomes a wandering minstrel taking brief excursions through his domain. If he encounters a rival male, he chirps more loudly. The chirps become longer, the song less rhythmic. If, on the other hand, his suit is successful, a female abandons her random searching for a mate and approaches him directly. In her presence his song changes gradually from a calling song to an ardent courtship song characterized by louder, many-pulsed, more rapid, and higher-pitched chirps. The nuances of these songs are lost upon human observers. We are aware only of change in loudness and tempo.

It seemed a pity to abduct any of these singers from their homes. On the other hand, if the secrets of their musicianship were to be revealed, there was no recourse other than to collect some males and, from the open field, some wandering females.

As I located each cricket I would clamp my net over it, firmly so that there were no openings for escape under the rim; then I would hold up the tail of the bag as an invitation to the captive to crawl upward. As he approached the narrow end I would coax him, or her, into the cul-de-sac, then work a small glass vial up to the end. Once the cricket was in the vial and escape closed off with a perforated cap, I was ready for the next catch.

I sorted my crickets in the laboratory. Into some cages went individual males, into others a pair of males, and to others a male and female. These assortments provided us with opportunities to study all types of songs—calling, aggression, courtship.

At one time, students of insect singing described songs phonetically—bzz bzz; sit, sit, sit; see see see see; and so forth. Unscientific as this may seem, it does strike a chord with English-speaking people. It would have to be modified to accommodate the pronunciations of other languages. Some entomologists have attempted to describe the songs by musical notation. Pierce's sophisticated instruments were able to resolve the song into its component parts, to measure the sound frequencies, to analyze acoustic details, and to gain insight to the mechanics of sound production. But the naturalist in the field must still rely on what his ear can detect. He can distinguish chirps, snaps, zips, buzzes, crackles, and the sequences of these that make up the "melodies."

Listening to this common insect without the aid of acoustic equipment is an ideal way for the novice student of insect music to tune his ear to this mode of music. Maturing as early in the season as they do, the field crickets have the stage to themselves. There are no competing performers in the same habitat in most of New England, so there is no chance for confusion. Furthermore, the song is relatively simple—chirp, chirp, chirp, one to three to the second, depending on the temperature. The situation is also simplified by the terrestrial habits of the males: the confusion of choral and polyphonic singing is absent. Early song rings out loudly and clearly against an acoustically neutral background.

One of the many things about these crickets that originally puzzled me was the complete disappearance in late July of their

songs, and of themselves as well, and their return after an intermission of a week or two. By that time there were many other singers abroad, but I had already committed the cricket song to memory and could pick it out from the background. Who were these singers?

August's field crickets were not a second generation. They were different singers performing the same repertoire. The crickets of early summer had sung their lays, courted, consummated their love, and died. Their consorts had laid eggs in the soft soil and died in turn.

For many years entomologists believed that the common field cricket had two broods each year. The spring brood was presumed to have overwintered as nymphs, which became adults in May and June. The fall brood was believed to have overwintered as eggs and become singing adults in August. Actually these two "broods" are two species, known respectively as the Northern Spring Field Cricket, which does indeed overwinter as a nymph, and the Northern Fall Field Cricket which sings until the hard frosts of October and November kill it. The following generation of this species overwinters as eggs. Only the time of appearance separates the two species. They are identical in physical characteristics, behavior, and song.

In late spring and early summer, there is only one other singer contemporaneous with the Northern Spring Field Cricket. He, too, has a twin fall relative. The Cox and Box scenario is played out again but in another setting and with a different kind of musical instrument.

The Spring Sulphur-Winged Locust

Sounds inharmonious in themselves and harsh,
Yet heard in scenes where peace forever reigns,
And only there, please highly for their sake.
William Cowper

*F*or several weeks now the spring field crickets had had the fields and roadsides to themselves. There were no competing singers. The frogs and toads were generally silent or, at most, lone voices in the woods. The robins and thrushes sang briefly in the darkness of early morning and the cool of evening. When I walked through the fields, my boots stirred up only the most minute creatures, diminutive ground crickets and young grasshoppers, some no larger than seeds of grass. It was with some surprise, therefore, that I encountered my second singer so early in the season, when all else seemed to be juvenile.

The meeting occurred just past mid-June on the first unpleasantly hot day of summer. The heat was so oppressive that the field crickets which had been chirping most of the night had ceased abruptly about ten o'clock in the morning and retired to the relative coolness of their burrows. The only flagrant signs of activity were displayed by butterflies. The first migrant monarchs had arrived from the south in time to savor the early blooming milkweeds. Dun-colored skippers rocketed erratically through and above tall grasses. Small blues and coppers minced about more gracefully. The orange Least Skipper, the smallest of its tribe, outnumbered all the others. An occasional Tiger Swallowtail strayed briefly from the treetops and sailed by in majestic hauteur, skimming the bushes. Yellow and green were still the dominant colors of the fields, but now the first daisies were beginning to accent that background with splashes of white, à la Monet.

Although the eye's need for stimulation was appeased, and the nose was tantalized by a green odor, the fragrance of freshness, there was nothing to delight the ear. With no thought or specific design in mind, I set off on the road leading down the valley and into Franklin. Since I did not anticipate finding any singing insects, I had left my net behind, thus breaking a cardinal rule of entomology. The road passed through patches of forest but for the most part traversed farmlands, areas where the vegetation was more or less uniform. The farms were, by design, monocultures. The forests, by the laws of ecological succession, were either dense stands of white pine or moderately mixed hardwoods.

In contrast, the edges of the road offered an embarrassment of riches, as do most borders. Roadsides, hedgerows, fence rows, railroad right-of-ways are everywhere the richest of habitats. Spared by plow, bulldozer, and ax, they are plant conservatories of astonishing diversity. They are heterogenous havens for insects, birds, and mammals that prefer diversity, Victorian extravagance, to the modern monocultures of agricultural lands, and esteem the cozy confines of brush, thickets, and patchy disorder in place of the freedom and exposure of the broad expanses of meadows.

I was not thinking of these matters as I ambled along, and I was thus unprepared for my meeting with the Spring Sulphur-Winged Locust. On one of the drier, more gravelly stretches of roadside I caught a glimpse of what appeared to be a yellow butterfly. By the time I could bring my attention to focus, the flash of yellow had dived to the ground where it seemed to have been swallowed up instantly. Slowly I walked over to the spot, a somewhat bare weedy patch of gravel. At my approach, there was again a flash of yellow, accompanied this time by a soft rattling. This unmusical tattoo carried its performer a distance of five feet. Again the ground received and hid the flyer. This time, however, I marked the spot well. I scrutinized it pebble by pebble. Finally the form of a locust detached itself from the background.

Locusts are easily distinguished from other Orthoptera. The distinguishing features are antennae and wings. The antennae of locusts and grasshoppers are short, compared to the long filamentous feelers of crickets. The wings of locusts are pitched tent-like over the back of the body, whereas those of crickets lie flat. There are other, more subtle distinguishing characteristics known to entomologists, but a glance at antennae and wings will suffice for the novice collector.

The locust that I was observing was rather slender as locusts go, and dark brown (some specimens, I learned later, are light brown and some almost black). The crest on his back (pronotum) was unbroken and looked like the crest on a medieval knight's helmet dented by the blow of a battle ax.

I sank slowly to my knees and extended my arm inch by inch. At the moment that I was ready to strike, a car appeared around a bend in the road, slowed as it approached, and came to a squeaky, dusty stop directly abreast of me. I was caught in an awkward, unconventional, inexplicable posture.

"Are you OK?"

The accent identified the speaker as a New Hampshireman.

Without changing position (I was now on all fours), I screwed my neck around to answer.

"Yes."

"You're sure."

"Yes."

"Can you get up?"

"Yes."

My laconic replies did not discourage my interlocutor in the least. Had it been later in the season he probably would not have stopped, because by that time news of the laboratory's existence had got around. The eccentricities of Professor Pierce and of his two young assistants had become known and taken for granted— not understood, but reckoned as harmless. It was, after all, a free country. Early in June, however, the identities of Paul Donaldson and myself were not common knowledge, nor was the nature of our activities.

There was a silence. I realized that I was still crouched in an unusual position, all by myself in the middle of nowhere, with my hand pointing toward space. I had not moved. I wanted that locust.

Finally, after many seconds, "What are you doing?"

On future occasions I learned to fabricate some patent but plausible nonsense. The truth only stimulated more incredulity. But this time I told the truth, stretching taxonomy a bit.

"I'm trying to catch a grasshopper."

"Goin' fishing?"

"No."

"Grasshoppers ain't good for fishing anyway. Wrong time of year, too. If it's grasshoppers you want, there's more in the pasture."

I nodded.

Finally, giving up in the face of my lack of civility, the driver shook his head in bewilderment or pity, let out the clutch, and chugged down the road. I could imagine the tale he would tell. Entomologists, amateur and professional alike, have often been the butt of humor.

Needless to say, the locust by this time had taken flight. Even had I been able to articulate a sensible answer to the driver's queries, the evidence had long since vanished.

Farther along the road I flushed another locust. This one I caught successfully by approaching from the front, so that as my hand moved forward he leaped into it. With gentle handling I was able to extend the wing cover on one side and then spread the pleated folds of the hind wing, as though opening a fan. The basal two-thirds were a striking sulphur-yellow, while the outer third sported a dusky curved band. Near the front edge of the wing the band sent a dark ray toward the wing articulation.

Having no container, an item that I would never again be without, I wrapped the locust loosely in my handkerchief, which I then stuffed, also loosely, into my pocket. I would be back at the laboratory before he could have time to find his way out of the cotton labyrinth.

Sulphurea, like the Northern Spring Field Cricket, had attained adulthood before all the other singing grasshoppers because it had passed the winter as a nymph rather than in the egg stage. Protected in detritus throughout the winter, it had gained a long head start over its grasshopper relatives. By late May or early June it had reached maturity and was ready to sing.

The Sulphur-Winged Locust has two songs and two musical instruments. When flushed from the ground it makes a brief (three- to four-foot) erratic circling flight and on a final abrupt turn plunges to the ground. During flight it produces a series of short sharp clicks, which can be heard from a distance of about thirty yards. This is the most common sound heard in early summer, as one walks along dry roadsides or through dry pastures and wastelands.

There is another song that can be heard only if one is patient and close to the locust. This is a series of weak, long, irregular buzzes that it makes by rubbing its legs rapidly against its wing covers. In the process, it appears to be stamping the ground, as though in a fit of frustration.

These two early summer musicians, the Spring Field Cricket and the Spring Sulphur-Winged Locust, between them exhibit the three kinds of song-making instruments characteristic of Orthoptera. The field crickets, in common with ground crickets, bush crickets, tree crickets, and katydids, produce sound by rubbing the wing covers together—file on one, scraper on the other. In flight, the Spring Sulphur-Winged Locust snaps and crackles by hitting the hind wings against the wing covers. It and its close relatives have, on their wings, a line of pegs that serve the same purpose as the files of crickets. Other locusts, not so closely related, bear the row of pegs on their legs.

It is just as well that the Sulphur-Winged Locust sings when the competition is absent, because his music has a baroque softness not designed for the vast concert hall of the grand open spaces. His style is more that of chamber music, and his instrument lacks the brilliance and forte of later singers.

Later in the season I was to meet his alter ego, the Autumn Yellow-Winged Locust, which is almost identical in appearance. His behavior is slightly different, however. His flight is longer and his song louder.

A Confusion of Ground Crickets

The cricket to the frog's bassoon
His shrillest time is keeping;
The sickle of yon setting moon
The meadow mist is reaping.
John Greenleaf Whittier

Throughout June, with only one species of cricket chirping and one species of locust crackling, there was no cacophony of song to confuse or challenge our acoustic exploration. The intensity of research that had characterized our introduction to the spring cricket and locust waned. There was nothing new to listen to and hence no work to be done. We were in the doldrums. To those engaged in scientific research, a letdown at the end of a phase of investigation is well known. The question of the moment had been posed, the mysteries resolved, the proposed solutions tested and achieved. Conan Doyle describes the condition insightfully when he tells about the apathy and irritation that Holmes feels once he has solved a crime. One might call the need to do something the "new-worlds-to-conquer syndrome." In our case, however, we could not create the new world; we had to wait till it came to us, when midsummer insects completed their development.

In the meantime, insects were fed, cages cleaned, records photographed and enlarged for analysis. Professor Pierce made the actual analyses. Paul busied himself with matters photographic. I burnished my education relative to Orthoptera by studying dried specimens, counting the number of teeth on wing files, watching male crickets courting and jousting. To while away the remaining time, I essayed a few minor experiments.

One of these was the "switched-wing experiment." I had no-

ticed earlier that the field crickets were right-winged; in all individuals that I examined, the right wing cover overlapped the left.
Thus, the file of the right wing was played upon by the scraper of
the left. I had also noticed that *each* wing had a scraper and a file
(a situation of which specialists were well aware). What, then, if
the overlapping positions of the wings were reversed? Would a
cricket rearrange them in the proper, and presumably more comfortable, position? If not, could he sing with the new arrangement?
The reversed-wing song was recognizable. But just as a right-
handed person writes only laboriously with his left hand and produces a crude script, so did the right-winged cricket fiddle crudely
with his wings reversed.

The experiment did not advance science in leaps and bounds,
and I learned later that years earlier another scientist had performed the same experiment. He, however, had switched the
wings of a cricket that had just become an adult. It had done quite
well; it had had nothing to unlearn.

A second experiment involved removing some teeth from the
middle of the file. This test did add to our knowledge of the mechanics of sound production and of the contribution of tooth-
impact frequency to the overall pitch of the song.

These digressions still left time to spare, and we occupied
that with extracurricular activity. Paul and I customarily drove into
town to have lunch and to run household errands for the Pierces.
We extended this interlude by exploring Franklin. There was only
one reasonable place to eat, the Franklin Café on the west side of
the river. Although the menu was not gourmet, it was appetizing
and, more important, within our modest means. Our standard
lunch was hamburg steak and a dessert that alternated between
gingerbread with whipped cream (real) and coffee jello with
whipped cream. As we soon established ourselves as regulars, the
Greek restauranteur and his two waitresses favored us with extra
helpings and side dishes gratis. During our entire stay in Franklin,
people in the café never came to accept our description of our

occupation. Eventually we were catalogued simply as handymen for one of the summer people on New Boston Road.

During lunch hour we shopped at the hardware store in the main part of town and at Surowiec's Market. Mr. Surowiec was puzzled by our routine purchase of more lettuce than two average-weight young fellows could reasonably be expected to consume. At first we refrained from explaining that the lettuce was for a large menagerie of voracious insects. But since we worked for Professor Pierce and the Pierces were highly esteemed customers we were finally accepted for what we were—harmless "bug catchers." Mr. Surowiec and his family were kind beyond measure. On more than one occasion we were invited to share family dinner and take part in special Polish celebrations.

After lunch we returned to the laboratory to do busy work: pasting records into books, tidying up, now and again helping with odd jobs around the house. We even played the role of "sidewalk foreman" at the site of a well that was being dug for the Pierces. It interested me that Professor Pierce, a learned physicist, should have engaged a dowser to locate water. He later resolved my doubts by explaining that he did not believe in water-dowsing but that he had great faith in the knowledge that these practitioners had of the lay of the land and local groundwater and ledge characteristics. I believed him. I even believed him when he proudly recounted stories of an uncle who had been hanged as a horse thief in Texas. Pierce had a quick wit, matched by a droll appearance that was the antithesis of the typical Texan's. He was short and rotund and attired most of the time in old riding breeches and high woolen socks.

At other times I strolled idly through the fields to acquaint myself with the area and, admittedly, for sheer aesthetic pleasure. In the process I stirred up hordes of grass-dwelling insects that flew, hopped, jumped, and scrambled to safety. Far and away the most common of these were legions of small black crickets, somewhat resembling the large field crickets.

Ground crickets

I picked up one of these miniatures for closer examination. It was only about four millimeters long, immature, and equipped with stubby wing pads instead of wings. It differed from an adult in only one other respect: its head was very large in proportion to its body.

Disproportionately large heads are characteristic of many young animals, human beings included; and there is something appealing about large young heads. They arouse feelings of tenderness. When they are combined with large eyes, the overall impression is one of daintiness, innocence, and vulnerability.

These musings fit well with the mood of the season, a mood of quiet unhurried growth, of promise rather than frenetic activity, of time stretching calmly toward fulfillment, the season in ascendancy. But a few days into the month, the mood and the laziness came to an end. Shortly before haying time, the fields exploded with song. With the passing hours, there was a crescendo of trilling and chirping. The air vibrated with continuous, pervasive, and totally enveloping song. The effect was exhilarating rather than offensive. Indeed, many people not tuned to the singing are unaware of it until it is called to their attention.

Thousands of young crickets had completed their final molts. In a single physiological burst, they had passed from infancy to maturity. Of course, internal development had been marching in a gradual ordered way, but in external appearance the change had been abrupt. Suddenly the crickets had wings with which to sing, and there were songs to be sung. Life was no longer exclusively a matter of nutrition and survival. Now it was nutrition, survival, courtship, and reproduction.

For us it all meant the beginning of work. These ground crickets, unlike their larger relatives the field crickets, were not restricted to well-spaced territories, nor did they sing like well-trained choruses. They sounded, in an insectan sort of way, like an enormous symphony orchestra tuning up before a concert, but in this instance the tuning up *was* the concert. Each insect musician played his instrument without any regard to others. He did not, however, have to tune it; his genes had done that for him. Once tuned, it did not vary.

The volume, mixture, and confusing variety of sound put us to the test. There was no way to isolate singers in the field, especially since they were generally tucked away in tangles of grasses. The work would have to be conducted in the laboratory. I therefore adopted the strategy of sweeping my net vigorously from side to side as I advanced across the field. I stopped only when the bottom was weighted with debris.

The challenges began when I got the mess back to the laboratory. The first challenge was to separate the wheat from the chaff, so to speak. Unless you have been confronted with a similar task, you have no idea of the bounty and complexity of a field—and the net's contents were only a small sample. Against a mass of botanical detritus and multitudinous creatures squirming, clawing, struggling, fighting to get to its surface, there emerged young crickets, mature crickets, small grasshopper nymphs, caterpillars, small flies, aphids and leafhoppers by the hundreds, parasitic wasps, spiders, here and there a honeybee, springtails, thrips, spittle bugs, and even a butterfly or two caught when the blossom on which they had been feeding was severed from the plant. This throng emerged from the compacted vegetation into which it had been tumbled: a botanical richness composed of blossoms, sepals, petals, broken stems, stalks, blades, and the seeds of myriad varieties of grasses—all dusted with pollen.

I soon learned that the place to sort this lively chaos was not in the laboratory but on the porch, where the winnowed sediment could be swept directly into the field. Timing was important in this endeavor, because emergence from the net was like a process of distillation in which the most volatile components of a mixture come off first and the least volatile last. The strong fliers escaped first, the crawling caterpillars and smallest, least agile walkers and hoppers last. The crickets were in the intermediate range of "volatility." By the time the fliers had departed, the crickets were walk-ing or hopping upward toward the rim. At this point I inverted a screened cage over the mouth of the net, thus allowing the crick-ets to incarcerate themselves. The maneuver was not perfect. Some of the more agile beetles and the laggardly flies joined the throng, and some crickets escaped. The former could be weeded out later. The escapees represented but a small fraction of the haul.

The next challenge came when the time arrived to study the singers and their songs. Even casual listening revealed that there were at least four kinds of songs. There appeared to be, however, only one kind of cricket. The question was: Did this cricket have a repertoire of four songs, or was I mistaken in believing that there was only one kind of cricket? One speaks of a murmuration of starlings and an exultation of larks, but the only apt designation for the tumultuous throng in the fields of July is a "confusion of crickets." The problem of sorting this out is analogous to that fac-ing a person who attends a symphony concert for the first time and wishes to know which instrument contributes what to the symphony, and which musician is playing what instrument. One must become acquainted with the tone and range of individual instruments and be able to identify each instrument by sight.

The first step toward a solution of the cricket problem was to isolate singers and record their songs for an extended period of time. This done, it soon became apparent that there were two basically different songs and that an individual insect sang one or the other, never both. One song was a trill, the other a series of chirps, weaker and dissimilar in intonation to that of field crickets. The trill consisted of twelve to nineteen pulses per second, be-coming more rapid as the air temperature increased. Although some were continuous, most trills lasted for two or three seconds and paused for varying periods of time. The trill was silvery, tin-kling, ti-ti-ti-ti or cree-cree-cree. This was the calling song. As with field crickets, the ground crickets also had courting songs. There were three or more rhythms to these songs.

The chirping individuals called at a rate of two to four chirps per second. Each chirp consisted of a train of four to twelve pulses, and, as is true of the field cricket's chirp, the pulses are imperceptible to the human observer. The rhythm of these songs also changes during courtship.

The third challenge I faced was that of distinguishing singers by sight. It had been faced before by every serious student of ground crickets. How many kinds of ground crickets were there? Which were species? Which were races? How many were merely variants of a single species? Coloration was an unreliable guide. Morphological differences, if any, were trivial and variable. As a consequence, the scientific names assigned to different crickets were constantly being changed. The most reliable guides were song and habitat. Not surprisingly, the only reliable structural guide for closely related species is the number of teeth in the file, but why bother with this when the song is so clearly diagnostic?

Judged on the basis of song and habitat, our triller and chirper turned out to be two of the more common species in the eastern part of the United States. The triller is Allard's Ground Cricket; the chirper, the Striped Ground Cricket. They occupy the same fields, lawns, and pastures, but the triller has a preference for the more well-drained locations, whereas the chirper likes the moister, more poorly drained grassy locations. By their songs and their habitats you may know them. Together they add richness to the acoustical dimensions of nature.

In those early days of summer the sounds of the gloaming, when the air is still and laden with moisture that blurs the visual dimensions of depth, provide an auditory perspective to distance. In the grass at one's feet the ground crickets trill the *here* while the chirp of a laggard field cricket marks the farther boundary of the field. In the forest beyond, muted but unmistakable, the song of a hermit thrush channels one's aural perspective to the purpling hills. There, as far as the ear can focus, is silence, the remote, the unperceived. In that remoteness is peace.

Climate in the Least Space

Respect all such as sing when all alone!
Robert Browning, Paracelsus, Part III

Although past records of the capture of various ground crickets reported them as frequenting different habitats, I did not appreciate subtleties of climatic conditions in these habitats until I found Allard's Ground Cricket frequenting slightly moister areas of a field than those preferred by the Striped Ground Cricket. I began to realize then that differences in habitat that were significant to crickets were not always appreciable by human beings, except in the grossest sense. Accordingly I began to search for local areas that might offer different climatic conditions. My first successes came as a result of my residing in town and having few diversions to occupy my spare time.

When I first arrived in Franklin, I rented a room in a lodging house not far from the railroad station. The location was favorable relative to the center of town, but the amenities were few. After the workday was over, there was little to do. The most momentous event every evening was the passage through the station of the daily fast freight from White River Junction in Vermont. Its coming was the focus of a gathering of old-timers. More local news, gossip, reminiscence, and rural philosophy changed minds here than could possibly have been accommodated by the town newspaper. Attendance provided me with opportunities to become acquainted and to converse with someone other than crickets. So long as I did more listening than talking, I was accepted at a cut above the kindly level accorded a well-mannered stray dog.

While listening to the conversations, I kept one ear more or less cocked for night noises. Later in the summer there would be many, but toward the end of July and out of earshot of the fields there were no sounds of interest. In fact the only sound was the mournful whistle of the approaching train echoing in the hills to the north. The sound was as different from the shrill atonal shriek of modern diesel locomotive whistles as a Chopin nocturne is from modern rock and roll.

As the whistling became louder, an old-timer would comment sagely on the train's progress. "She's just passed Jim's place. She'll be here in three minutes." Watches were pulled out of pockets. In four minutes the train arrived, first the beam of the headlight, then the full-throated whistle, then the locomotive thundering past in clouds of smoke and steam. It was followed by the prolonged clicking, clattering, banging, and groaning of a hundred or so freight cars. As the caboose flashed by and its red light disappeared in the darkness, the gathering broke up with assurances of reassembly the next evening.

My lodging did have one advantage: it placed me midway between Webster Lake and the laboratory. The former was an attractive place for weekend swimming. More to the point, it introduced me to a new cricket. Journeying toward the laboratory took me past a gravel pit, where I found still another species of ground cricket. Both experiences sharpened my appreciation of microclimates, picturesquely and aptly called "climate in the least space."

The circumstances were these. There was a sandy beach at one end of Webster Lake. At most other stretches along the shore, mixed oak-hickory-pine woodlands came down to the water. One warm Sunday in the first week in August, I had extended my swim beyond the limits of the beach and was floating not far offshore. Across the lake a dog barked petulantly. Low-keyed voices from the beach contributed the only other sound. In those days there were few if any outboard motors, and no water skiers being towed noisily back and forth. The immediate vicinity of the lake was quiet, because there were no fields of singing crickets close by.

It was during this interlude of floating that I heard a few

crickets along the wooded shore. The sound surprised me, because with my limited knowledge of crickets I had always associated them with open fields. Clad only in swim trunks I was not ideally attired to investigate a brushy shoreline, so I decided to return the following day.

The next day, equipped with net and vials, I approached the spot from the shoreward side. There were indeed a few crickets trilling. The songs were new to me. They were high-pitched, bell-like tinkling songs distinctly different from the trills of Allard's Ground Cricket. The singers were concealed in the litter of oak leaves, but I managed after several unsuccessful tries to capture four.

These crickets have been aptly named Tinkling Ground Crickets. Their calling song is almost identical to that of Allard's Ground Cricket, with one important discernible difference: the intervals between chirps are longer. It is this slower frequency, five to ten chirps per second, that makes the song less of a trill and more of a tinkle. Also, the pitch is slightly lower. The crickets themselves are almost indistinguishable insofar as coloration and structure are concerned. Once you get to know the tinklers, however, it is possible to discern a subtle difference, but it is an impression that is almost impossible to express in words. You know that you know, but you do not know how you know. You can count the teeth on the file (they are reliably diagnostic), but it is much simpler and equally reliable to learn the song and the habitat.

The habitat, leaf litter at the edges of hickory-oak-pine forests, is a most reliable guide to this species. The selection by the Tinkling Ground Cricket, as with the selections by other species (well-drained grassy fields, poorly drained grassy fields, and gravel pits) raises interesting questions concerning crickets' choices of homes. The crickets usually stay reasonably close to where they were hatched, so in that sense they do not choose their habitats; but if they stray out of their neighborhoods, they tend to return. The three species to which we have been introduced thus far may live within a few hundred yards of each other but rarely intermingle. I found Allard's and the Striped Ground Cricket intermingling where their respective neighborhoods met, but I never found the Tinkling Ground Cricket outside its dry leaf litter territory.

Although we large creatures can distinguish the habitats only grossly, the crickets are more subtly tuned to microclimatic differences. Crickets are smaller, live closer to the ground, and are unaware for the most part of the broad climatic differences that are so obvious to us. The climate in the least space is generally more stable. In the upper atmosphere at the level of *our* heads, contrasting conditions—that is, cold air, hot air, dry air, wet air—are almost immediately equalized by free movement of the air. Near the ground, on the other hand, contrasting conditions exist side by side. Mixing and equalization are impeded by friction acting on air as it moves over the ground, and modified by the kind of soil, its form, the plants present, the obstruction that these impose on wind movement, and a crazy-quilt pattern of shade and sunshine. Plants change the amount of radiant energy that reaches the ground from the sun in the daytime and the amount of outgoing radiation at night. The plants themselves have their own temperature and water economies, and their effect on the microclimate changes as they change in form, size, and number.

All these subtle and complex physical interactions shape the climate of the world in which ground-dwelling insects and other inhabitants of low vegetation live. Crickets and other small insects are particularly vulnerable to climatic changes in their microworlds. Since they are small creatures, in which the surface of the body is large compared to the volume, loss of water is an ever-present danger for them. And because they cannot regulate their body temperature, they are easily subject to overheating.

Each species has its own adaptation to a particular microclimate. Thus, Allard's Cricket is best adapted to moist grassy places, and the Tinkling Cricket to dry sunny/shady places. Even within these small worlds the crickets encounter contrasts. At the surface of a leaf there is rapid heat exchange, such that a thin zone of air, about ten millimeters thick, is at a temperature different from the surrounding air. The upper surface of a leaf may be warm or cool depending upon its rate of transpiration, its size, shape, reflectance, texture, and height above the ground. The under surface of a leaf is usually cooler and more humid than the upper. Insects are sensitive to these nuances of climate and take advantage of them. When cold, they may bask in the warmth of leaf upper surfaces; when overheated or dry, they may seek the cool, humid environment of under surfaces. Different species of insects have their own characteristic responses. They frequent a given plant at different times of day or visit different plants.

Where vegetation is sparse, temperature extremes are great, low humidity is a problem, and exposure to weather and predators is considerable. These were the conditions that I encountered at my next stop, the gravel pit.

The pit was located about midway between my lodging and the laboratory. It was not especially appealing, except perhaps to bank swallows, geologists, and sand and gravel contractors. In this part of New England, glacial gravel pits are very much alike. Glaciers are untidy when they abandon an area; they are great geologic litterers. In their wake they leave rocks, boulders, pebbles, gravel, and sand, heaped willy-nilly into moraines, eskers, drumlins, and ancient meltwater stream beds. Leaching of organic nutrients from the soil had left the pit dry and botanically unproductive. Once a drumlin or moraine had been exploited for its sand and gravel, conditions were not felicitous for any but the hardiest plants. Exposed sandy ground absorbs heat rapidly during the day but retains it poorly at night, so near the surface such ground tends to be hot in the daytime and cool at night.

Having become acquainted with ground crickets in fields and forest borders, I did not think I would find any living in the gravel pit; nevertheless, I wanted to make a cursory exploration on my way to lunch. I expected some Sulphur-Winged Locusts and perhaps related species, but I hardly expected to find crickets. I was therefore surprised to hear trilling.

The composer of the song was a dull reddish-brown cricket clothed in gray pile. His overall appearance was sandy, so much so that when he was still he could hardly be picked out from the background. Of all the ground crickets, the Sand Ground Cricket was the only one which could be identified immediately, without hesitation, and without benefit of song. The trill is high-pitched, like those of the Striped and Allard's Ground Crickets, but faster. Trills occur at a rate of twenty-six to thirty-seven chirps per second.

There was one other addition to our song collection as a consequence of my living in town. The circumstances illustrate the value of knowing songs well enough by ear (in contrast to understanding technical details). They also illustrate some of the hazards to which an entomologist is exposed.

One night I was returning from a late movie, and as I passed the local graveyard, I heard some crickets trilling continuously but with an unusual pulsing, as though the singers were stuttering. Obviously no collecting could be conducted in the dark. The next night, however, I returned with a flashlight and some glass vials but no net. I would have felt self-conscious wandering through town, day or night, with an insect net in hand; but I was destined to be embarrassed anyway.

The crickets seemed to have congregated in dense un-trimmed grass around headstones. I crawled among the stones, switching the flashlight on whenever I thought a cricket was nearby. The search had progressed for half an hour and I had managed to capture two crickets by hand, when out of the stillness of the night a gruff voice above and behind growled, "Hey you! What the hell do you think you're doing?"

I jerked around into a blinding light. Another voice added, "I think we've caught one of them gravestone vandals."

"Get up, you!" commanded the other voice.

I was grabbed forcibly from behind.

It was obvious what had happened. Someone had seen a light moving mysteriously around in the graveyard and had alerted the sheriff. He and his deputy had arrived quietly—and found me.

Eventually, after prolonged explaining, presentation of identification, and a call to Professor Pierce, the sheriff released me, with stern admonitions about watching my step.

The fruits of my embarrassment were specimens of the Carolina Ground Cricket. He was as indistinguishable from his relatives as they were from one another—except, of course, by specialists. His song and his habitat were his marks of distinction. His habitat is low, densely grassed borders of walls, gardens, and swampy woods, where he lives in matted herbage and beneath stones and logs. He trills continuously with considerable volume both day and night. The trill is lower in frequency than that of his relatives in the area but has the highest rate of trilling: sixty-eight to seventy chirps per second. It is, however, the stuttering, pulsating nature of the average song that is most clearly diagnostic to the ear.

Together with Allard's Ground Cricket, the Carolina Ground Cricket is the most common and widespread of the group. Having identified his song and become acquainted with him, I was able to locate other specimens without being mistaken for a vandal or the modern-day equivalent of the eighteenth-century surgeon John Hunter, who robbed graves to acquire bodies to study. *Sic transit gloria scientiae.*

The Sphagnum Cricket

How dreary—to be—Somebody!
How public—like a Frog—
To tell your name—the livelong June—
To an admiring Bog!
Emily Dickinson

Most often my discovery of a new cricket soloist emerged from a carefully planned strategy. After consulting the "bible," Morse's manual, I would visit a particular habitat at a specific time of day and season, find a comfortable spot to sit (if the terrain permitted), relax quietly, and listen. I searched with my ears. This strategy demanded single-minded concentration. Other times a new discovery was entirely serendipitous. On those occasions a serenity, a contemplative mood, even a dreamy detachment from my surroundings played an important role because any unknown song penetrated the consciousness as a disturbance, an interruption of the mood.

There is a feature of interruption that seems contradictory: interruption of a flow of sensory information increases rather than decreases information to the brain. The contradiction comes about because of an apparent conflict between an animal's needs and its limitations. A pressing problem facing all organisms is that of maintaining a stable internal environment while exposed to an essentially alien external environment that is ceaselessly changing. We, for example, must keep our body temperature constant summer and winter; we must keep our blood sugar and salt levels constant; our heart must beat steadily. To sustain this tenuous balance, animals must monitor their internal milieu constantly and must also receive continuous and reliable information about the external world so that change can be counteracted. The realization of this goal is impeded by the energy limitations of sense organs; they tend to slow down and shut down under constant and unchanging stimulation. Even the brain stops "listening" to steady, monotonous signals. The solution to this problem is to interrupt constant stimulation or constant silence. Any sudden interruption, whether to silence or to a steady stimulation, jerks one immediately to attention. Thus it was that by interrupting silence and serenity the Sphagnum Cricket revealed his presence.

It all began with Professor Pierce's desire to dam the small brook at the edge of the forest and stock it with trout to be fished out at his leisure. By the end of a particularly wet July, the desire had become reality. The dam completed, the brook had provided a modest-sized pool which, as it turned out, served a variety of pleasures—a home for trout, fishing for the professor, a site for an occasional pause in my day's hunting for insects in sunny fields and pastures, and even a brief surreptitious swim on oppressively hot days, especially the dog days of August.

The pool was serenity, a stasis in the incessant activity of the brook, a contrast. The brook even in its quieter phases was the essence of motion. Not even the small languid pools that beaded its course were ever still. In the shade, one's eyes could focus on reflections of the tree canopy and sky. At the same spot but at another level of focus, small bits of flotsam appeared on the picture of the sky. At still another level of focus, behind the mirror, the bottom shone clear in a shaft of sunlight, as if there were no water. At other places, in the shade beyond the boundaries of the pool, water trickled sinuously over algae-covered stones and gravel until it reached a sunny spot, where it sparkled. In sunny stretches, water striders held their positions in the stream; each created four silver spots where four of its six legs dimpled the surface. If I focused my eyes on the bottom, I saw the shadows of the dimples moving across the sandy shoals in synchrony with the silver dimples above. Occasionally a small frog perturbed the scene by plopping from the bank. The banks were clothed in various species of ferns—Cinnamon, Sensitive, Lady, Interrupted—rising from cushions of moss and sphagnum. The artificial pool, by contrast, seemed preternaturally placid. Whereas the brook stimulated, the pool hypnotized. One's mind was lulled to dreamlessness.

Recollection of these scenes tempted me one day in August while I was hunting for Ledge Locusts and Snapping Locusts on the hot ledges of outcroppings in the upper pastures. I yielded to temptation and made my way to the pool. But my desire to swim went unfulfilled because Professor Pierce was sitting on the rustic bench near the water, busily tying a fly. Observing him was Professor Cannon. Obviously I could not swim, nor could I retreat gracefully, since I was supposed to be elsewhere.

Without looking up from his task, Professor Pierce kindly invited me to tarry a while. I suspected that he knew quite well why I had come to the pool, but he gave no hint either by word or change of expression.

The two men had not been conversing, else I would have heard them before I emerged from the forest. The silence which had been broken only by my arrival was complete. No birds were singing. The brook below the dam was quiet; since a deluge in July, there had been no rain to send more than a thin satin flow over the spillway. There were no breezes. Not even the trembling aspens stirred. The pool itself was a dark mirror flawed only by the transient footprints of a water strider and expanding rings of bright water inscribed by some whirligig beetles.

It was this silence that was interrupted: I thought I heard a feeble trilling coming from a great distance. So faint and unfocused was it that I was uncertain whether it was reality, imagination, or a case of tinnitus. Sometimes silence is so complete that you begin to hear sounds from your own body, your breathing,

Professors Cannon and Pierce at the pool

your heartbeat. When your ears eventually adapt to the monotony of these, your mind, wearying of perfect silence, creates illusions of sound.

Was the trilling an illusion? By an intense effort of concentration I convinced myself not only that the song was real but that it came from low-lying ground not far beyond the edge of the pool. That ground was a bog of pastel gray-green sphagnum moss accentuated here and there with scattered low sedges and rushes and an occasional clump of sheep laurel.

One does not crawl on hands and knees in a cold, wet, spongy sphagnum bog. Before taking even one step, a prudent stalker ranges in by ear on the source of the sound. Then, taking care not to cause the springy bog to quake, one advances as slowly as a heron stalking a fish. In this manner I succeeded in approaching within about ten feet of the spot whence I calculated the sound to be coming.

So far so good; the trilling continued. Again I stole forward, mouth agape, holding my breath. I do not know why one tends to hold one's mouth open when listening intently. I have been told that this unconscious reaction intensifies the acuity of hearing. Perhaps. True or not I was not conscious of anything except the trilling, which now seemed almost underfoot. Scanning the ground ahead, I took one more step. The trilling ceased. For perhaps five minutes I became a statue. The hunt had become a waiting game. As I would find on so many occasions, the pursued seems to have more patience than the pursuer. Some crickets in particular are possessed of annoying reserves of patience.

The trilling resumed. I could see nothing on the sphagnum around me. I took one more step. Silence. Finally, convinced that the singers had hidden themselves, I forgot about getting wet, knelt in the bog and pressed down the sphagnum. Like a giant sponge being squeezed, it gave up its water. My weight and my squeezing produced a small puddle. Then I saw them—small, shiny crickets trying to jump from the surface of the water to safety and concealment. They were the smallest crickets I had ever seen, the smallest of all ground crickets, as I was to learn later. Quickly laying my open net on the ground, I began to press down the sphagnum all around it, making a larger puddle. Frantically the small ones swam, hopped, crawled, into the net.

Without bothering to examine my catch, I herded the throng toward the bottom of the bag and choked off the opening. The most direct route back to the laboratory was past the pool. I headed that way.

"Caught something?" asked the professor without taking the cigar stub from his mouth.

"Uh, huh!"

"New?"

"I think so."

"Humph! You've done better than I'm doing. Did you know they were down here?"

He looked at me poker-faced. I did not rise to the bait. He knew as well as I that I had come down for a swim.

"No."

He kept looking at me over the top of his glasses. Throughout this laconic exchange Professor Cannon said nothing, but I thought I detected a suspicion of a smile.

"I'll take them back to the lab and find out what they are."

Back at the laboratory I transferred my catch to one of the standard cages. They were indeed midgets—only about five millimeters from head to tail. At first glance they seemed to be glittering fragments of jet. On closer examination they looked like miniature field crickets. The body plan was essentially the same, rounded heads, flat wing covers, antennae longer than they themselves. There were both males and females in the group. That was all to the good, because companionship would bring out the best and most varied musicianship in the males. In the meantime they would settle themselves in various parts of the cage and use the lettuce that I supplied both as food and concealment. The next day Professor Pierce would focus his microphone on them and record their songs.

The song, heard at close quarters, still retained the faintness and delicacy that characterized it when heard from a distance in the bog. It has been described as "feeble," but that characterization does not capture the fragile beauty and sprightliness of it. The song, arising from the silence and the sense of mystery embracing bogs, conjures up visions of fairy folk, even of elves and leprechauns, which are said to haunt these places. It is by their small size, their trill, and their habitat among the sphagnum and the will-o'-the-wisps that one identifies these crickets. There in the bogs these small creatures fiddle until the hoarfrosts of November silver the fields and the fallen leaves of the swamp maples lay a coverlet of red over the sphagnum, and the forest is still.

Shuffling, Rustling, Crackling

The russet grasshopper at times is heard,
Snapping his many wings, as half he flies,
Half hovers in the air.
Carlos Wilcox

Samuel H. Scudder, a nineteenth-century entomologist who pioneered in the study of North American butterflies and grasshoppers, wrote, "Crickets shrill and creak, longhorns scratch and scrape, shorthorns shuffle, rustle, and crackle." His onomatopoetic descriptions are helpful diagnostic expressions of differences between crickets, katydids, and locusts and supplement erudite scientific distinctions. His words do more, however, than set criteria for identification; when they are spoken aloud, they stir visions of crickets chirping in fields and gardens, of bush katydids in pastures overgrown with goldenrod, sawing away on their reedy ill-tuned violins, of locusts fluttering in crazy flight over barrens and ledges.

Of these orthopteran assemblages, locusts seem to be the least generously endowed by nature and the least esteemed by humanity. They are the creatures of biblical plagues and in medieval times were the victims of excommunication by the Church. They are the least musical of all the singing Orthoptera; indeed, most are mute. They are generally clad in dun, peasant colors.

Yet no creatures should be diminished in our eyes simply because their virtues are not ours. The mere fact that locusts are among the most ancient of insects, as are others of their tribe, having behind them a lineage as impressive in its antiquity as that of the dinosaurs, is alone reason for looking for the best in them. Not the least of their accomplishments is a tenure on earth exceeding the period of occupation by human beings and an ability to compete successfully with people. They flourish in barren places; they survive in desert heat and desiccating winds; their rickety flight carries them across continents.

There are some among them which, conforming to all solid locustan values, are exceptions in ways that bring them to our attention. They may not qualify as master singers, but they excel in aerial acrobatics and in displays of flamboyant colors. These nonconformists are the band-winged locusts, unmatched in color and versatility of flight by any other orthopterans. The field and ground crickets, in funereal black, chirp lugubrious songs of love; the katydids squawk and click while concealing themselves in unrelieved Lincoln green among blades of grass and amid green seas of leaves in herbage and forest; but locusts flaunt their colors and their flighty pretensions. Though their tune is feeble, their gay raiment is a reminder that the denial of one talent is often compensated for by the bestowal of another. Think only of the melodies of the modestly attired lark and nightingale and the parody of song emitted by brilliantly plumed cardinals and bluejays.

By early June we had already met one of these locusts, the Spring Sulphur-Winged Locust, a singer of limited accomplishment but an aerialist of inventive skill and so colorful as often to be mistaken for the Clouded Sulphur Butterfly. Now, in July, we made the acquaintance of the Black-Winged Locust, a conspicuously large locust with deep black underwings trimmed with yellow. He, too, was a reminder of a conspicuous butterfly, the Mourning Cloak.

As so often happens with unexpected events, he first appeared to Paul and me at an inopportune moment. Having just returned from a noontime shopping trip, we were carrying armfuls of groceries to the house. As we walked across the dirt driveway, we were startled by a splendid locust rising from the ground ahead. He ascended vertically to a height of about three feet, hovered there for a few seconds, then with an increasingly great thrashing of wings continued upward to a height of ten feet. There he hovered briefly, before gyrating awkwardly in a fall to earth like a game bird shot from the sky. He landed close to the spot from which he had launched himself.

During this remarkable performance his hind wings were displayed in full color; at low altitude they emitted a soft fluttering and higher up a muffled crackling. When he landed, he became silent, colorless, and motionless. His hind wings were tucked back under drab wing covers, which camouflaged him so well that he became almost invisible against the dirt.

His behavior and camouflage were beautifully coordinated; he remained motionless and would not be flushed. How nature achieves such rational coordination in a creature with a brain smaller than the head of a pin is one of the wonders of life. In this case, misalliance between form and function would have led to extinction. The congruency is still another cause for wonder at the organized complexities of nature and the strictures imposed by earth's inimical environments. There is no room for chaos, and only those creatures in whom genetics and development are in tune with place, time, and circumstances successfully populate the planet.

Knowing that the Black-Winged Locust was now abroad, we began that same afternoon to search for more. We spotted three in a patch not far from the laboratory. Obviously there was no point in bringing one into the building; he could not perform in a confined space. Professor Pierce could, however, bring the laboratory to him. We wheeled out the mobile receiving and recording apparatus that had been so useful before.

When this apparatus was positioned cautiously near the performing locust and the horn pointed toward him, a record of his sound—one could hardly dignify it by calling it a song—was obtained. At the beginning of hovering, the wings fluttered rapidly, not quite silently, for a period of little more than a second. As the locust rose he produced a succession of slower, louder clacking

The Black-Winged Locust

lasting from four to six seconds. This noise arose from the impact of the hind wings on the following edge of the wing covers at a rate of twelve per second. The initiation of sound was not an inevitable consequence of the physics of flight. The locust clearly controlled the onset and cessation of clacking.

Ability to manipulate the wings for purposes other than flight was the only musical accomplishment of which the locust was capable and, compared to what crickets could do, it was a paltry skill. Captured individuals which we observed with great patience in the laboratory made no sound whatsoever. Unlike Sulphur-Winged Locusts, which could produce small buzzes by rubbing their legs against their wing covers, the Black-Winged Locust when grounded was mute. He had neither file nor scraper on legs or wings.

After a few days during which he never made a sound (but consumed large quantities of fresh lettuce), we released him on the nearest patch of bare ground where he could indulge his small pleasures. For the remainder of the season a locust remained in the vicinity and, if the sun shone brightly, performed daily. I like to believe that this locust was our liberated captive, but I lack proof because I had neglected to mark him.

Encouraged by our successful recording of the Black-Winged Locust, I decided to explore dry, rocky, barren places in search of related species. Till this moment, collecting forays had been restricted to meadows and bordering woodlands. The greenness, luxuriance, dampness, and fragrance of those places seemed most in keeping with notions of fertility, bounty, diversity, and all the bucolic impressions that one has about summer. Moreover, all the singing had been coming from these lowlands. Having discovered the Sulphur-Winged Locust in June along barren roadsides, I should have been more open-minded about xeric areas, but it required the performance of the Black-Winged Locust to shake me from my romantic prejudice. Now was the time to set matters right.

The laboratory, as I have said, was situated on a low hill sloping toward the bogs and woods. On one side there was a working pasture in which later in the season the cows supplemented their diet with apples from the trees along the fence line. The apples, Astrachans, dropped early and lay in winey evanescence. Field and ground crickets gathered and tippled, as did some yellow jackets; and the cows got gloriously drunk. I shudder to think of what might have happened to the milk. On another side of the laboratory a hayfield, soon to be mowed, extended to the road. A ring of hills encircled the whole domain, some forested to the crest, others with abandoned pastures, bare and craggy. It was to these last that I now wandered.

The closest high pasture was old but had not succumbed to the encroachment of the forest. Most old fields and pastures, if left to themselves, would gradually return to their primeval forested states. The grass would be shaded out by the goldenrod and asters, these in turn by blueberry, bayberry, and bracken. Soon in the shelter of the hardy shrubs pine, birch, and poplar seedlings would gain hold, eventually to shade out of existence the very plants that had offered them shelter in the first place.

Often the soil was impoverished and so thin that even the rich bequests of generations of cows were leached away. Then the pastures, though released from the pressure of grazing, retained the characteristics that they had acquired when the land first was cleared. The place that I had chosen to explore had a few grassy patches but was mostly rocky and fit only for hardy vegetation that made few demands upon it. Granite ledges, those near the summit being impressively high and showing scars from glaciation, rose above massed bayberry, sweet fern, and bracken. Here on the ledges and the areas surrounding them, I met three relatives of the Black-Winged Locust: the Cracker Locust, the Ledge Locust, and the Clouded Locust.

I had not yet reached the top of the pasture when I heard the Cracker Locust on the rocks above me. There was no mistaking him. He is the loudest of locusts, but he is not the easiest to see once he has landed. This one had landed on a bare crag at the very top of the hill, a rock so exposed to the elements that lichens covered only its sheltered faces. The color of the exposed surface consisted of mixed shades of white, gray, and black toned with patches of rust and concentrations of hornblende. Against this tessellated background the locust was invisible. I must have ventured too close, however, because he leaped again into the air and took off at breakneck speed across the jumbled rocks, all the time emitting a loud firecracker staccato of snaps.

Time and again I tried to catch him. Each time he circled back to his starting point, so that he never abandoned the ledge. Finally he made the mistake of landing where I was standing motionless. I could now distinguish his slate and black outline. I caught him.

Farther along the crest of the hill, on another high crag that by some accident of wind currents, vortices, and humidity was more heavily encrusted with lichens, I flushed a Ledge Locust. Lighter in color than the Cracker Locust, he tended more to grayish-brown. His mottled wing covers matched the ash-gray of weathered rock as well as the greenish-gray of some lichens and the black of others. When he was not basking in the sun, exposed but invisible, he ventured on silent flights. When flushed, however, or taking longer flights, he rattled in typical locust fashion. Later, when I examined him in the laboratory, I discovered that he, like the Sulphur-Winged and Cracker Locusts, could make weak buzzing sounds by rubbing his legs against his wing covers.

As I descended from the high pasture, I crossed a sandy terrace which seemed to have been formed from sediments that spring freshets together with channeled runoff from the ledges had deposited there. The water had puddled long enough to unburden itself of silt but not long enough to establish an upland meadow. As a consequence, the terrace in summertime remained dry and sterile. Its only vegetation consisted of clumps of bunch grass.

I noticed in my passage among the clumps a most peculiar

Bunch-Grass Locusts

locust, at once graceful and grotesque. He had an acutely retreating thin face, a long slim body only partly covered by wings too short to fit well, and unusually long, slender hind legs. He was uniformly pale gray. In both color and body form he was superbly designed to resemble bunch-grass stems. At my approach he neither flew nor jumped away. Instead he sidled around a grass stem, keeping it all the time between him and me. This strategy might have worked with some predators but not the human one. He, too, was bottled for future study. Since he did not startle readily, I caught him so easily that it did not seem sportsmanlike. Later, when his songs were recorded, they were found to be weak buzzes lasting about three seconds. Each consisted of series of trains of pulses.

Farther down in my descent I passed through a zone of upland meadow, a habitat intermediate in character between the dry rock pasture and the more verdant humid fields around the laboratory. As I swung along through the timothy, I heard a buzz to which I paid scant attention because it sounded somewhat like a bumblebee, though louder and less tuneful. It also sounded rather like a droning beetle, and I was reminded of lines from Gray's *Elegy*.

> Now fades the glimmering landscape on the sight,
> And all the air a solemn stillness holds.
> Save where the beetle wheels his droning flight,
> And drowsy tinklings lull the distant folds.

The time, however, was not evening. I saw no droning beetle, nor did I see any bumblebees. With some effort I cast off my reveries and woolgathering in time to notice a locust flying across a sunny area of short grass. Compared to other locusts, this one was an easy catch. I did not have to risk my neck leaping pell-mell over uneven granite blocks.

When I finally caught the droning flyer, I saw at a glance that he was no beauty. Short, robust, somewhere between rusty yellowish and sooty brown, with different shades of mottling and barred legs, he was the anticlimax of the day. In the laboratory, however, he surprised me. He produced the expected buzz by rubbing his legs against his wing covers. But on other occasions he executed a duller, softer tattoo. In performing this, he looked a bit ridiculous: he stamped his feet against the floor of the cage, as though in complete frustration. We picked up the sound by placing the microphone of the recording apparatus directly against the floor. Thus we discovered that this musician played three instruments.

All these additions to our library of locust songs we owed to our introduction to the Black-Winged Locust. And there were more to come, now that my ears and consciousness were tuned to the elusive sounds made by locusts. Furthermore, my expectations were expanded by the realization that each species of locust, like each species of cricket, was to be found in a special habitat. These habitats extended in all degrees, from the most barren to the most verdant. Having thoroughly explored the barren, I began to listen anew in a more discriminating fashion in the fields and meadows. In short, I made a conscious effort to blot out the louder songs of the crickets and listen for the muted instruments of the orchestra.

To hear less obvious sounds, one must studiously acquire the ability to blot out sounds selectively. Some situations are more easily dissected than others. In a roomful of conversations, for example, a person can quickly hear his own name above the general chatter. Just as one can focus attention visually on a particular feature of a scene to the exclusion of the rest of the view, so can one focus aurally on a single sound to the exclusion of others. By such an exercise of attention focusing I discovered two more locusts in the field, one in the damper parts of the meadow, the other in the drier areas along an old rail fence and the border of the woods.

The inhabitant of the damp meadow, a lover also of moist ditches and grassy banks of brooks, was the Marsh Meadow Locust. He is a greenish-yellow grasshopper distinguished from others by a black bar extending back from each eye. He sings by rubbing his legs against his wings, but unlike the band-winged locusts, the file is on the legs and the scraper on the wings. Motion being relative, it makes little difference whether the file is drawn across the scraper or the scraper across the file. Differences in song depend more on the number and spacing of teeth, the rate of scraping, and the resonant characteristics of the sounding board—that is, the wings.

The song of the Marsh Meadow Locust is a series of soft lisping syllables. No one has described the sound better than Piers, a chronicler of the Orthoptera of Nova Scotia: "It is a soft, dreaming, lulling sound of the country characteristic of a quiet hot forenoon in August and September."

The inhabitant of the drier edges of the field was the Sprinkled Locust, a wood-brown species with shiny black side lobes on his thorax. He and his mate are especially partial to the vicinity of rail fences, old stumps, and bits of decaying wood. Females bore holes in these, as chambers for their eggs.

These locusts have the reputation of being the best jumpers of all Orthoptera. They jump powerfully, rapidly, successively when pursued. Not for them is escape by flight, by freezing in camouflaged paralysis, or by sidling around stems. They flee, settle down, and call to the female by rubbing pegs on their legs against scrapers on their wings. The sound is a subdued tsikk-tsikk-tsikk-tsikk.

There were other less common locusts to be found in the area; but with what we had, our week had been a busy one. We had added seven new singers to the insectan orchestra. We had become conscious of the advance and changing mood of the summer.

Meadow Grasshoppers

A thousand golden sheaves were lying there,
Shining and still, but not for long to stay—
As if a thousand girls with golden hair
Might rise from where they slept and go away.
Edwin Arlington Robinson, "The Sheaves"

On the morning of the first day of August, the stillness was shattered by an intrusive noise from the direction of the driveway above the laboratory. There were grating sounds of iron wheels crunching gravel, along with metallic squeaks and rattles. As the racket grew louder I could hear the clomping of heavy hooves. There was but one explanation: haying time had arrived and someone was approaching with a mowing machine. Two images came to mind. The first was a vision of another time when, in the last summer of my high school years, I had been a hired hand on a dairy farm in another part of New Hampshire. The second was a vision of stubble in a mown field.

The recollection of my summer as a hired hand conjured up images of the hot, dirty, sweaty aspects of haying. I had not been allowed to drive the mowing machine or hayrake with which the horse did all the work. My job involved pitching hay into the wagon, and subsequently, on arrival at the barn, climbing into the stifling hot loft. There, as each load was hoisted from the wagon and dumped in a cloud of dust and chaff, I forked and stacked it. This was my firsthand experience with the backbreaking, dirty toil of farming which Wordsworth and other poets of country life seldom eulogized.

My second thoughts dwelled on the ecological changes that were about to take place in the surrounding fields. These would have a direct impact on our research. The tall waving grass would be flattened and raked into windrows. Whatever shelter and concealment it had offered would be destroyed, and the field would become for a time hotter and drier. What would happen to the singers of high summer that were just making their debuts? We were fast approaching the climax of the insectan symphony.

Since the balmy days of June, more musicians had gradually been added to the orchestra. The music had already developed a richness, complexity, and change of mood. New sounds, tempi, and sequences were being introduced. Admittedly, musical analogies are stretched because in the world of insects each instrument is a primitive device tuned to a single pitch. On the other hand, there are many kinds of instruments and thus many tones in the composition. There are also, when all things are considered, passages forte and pianissimo, choruses, solos, variations on themes. There are no melodies as such, no composer to orchestrate the whole, and no conductor to interpret and direct, but the ultimate expression is a paean of nature.

Despite all the analogical contradictions, the effect is neither cacophony nor the "white noise" of physicists. The field as a whole projects a composition, almost a tone poem, that reflects successively the advent of summer, its fulfillment, its passing. One can listen to the field, and the woods as well, and so mark the days of summer.

I had been subliminally aware of the changes that were taking place, and, based on my reading, I had been anticipating what was yet to come as summer passed into fall. We had been so completely occupied with studies still in progress that excursions into the maturing field had been postponed. There was always tomorrow. Now the arrival of the mowing machine had announced that tomorrow was today. I went out to face reality.

Mr. Russell, the neighboring farmer, reined in the horse as I came to greet him. He was a jovial, sandy-haired, red-faced man in a gray shirt and blue overalls. He seemed to overflow uncomfortably the dish-shaped iron seat of the mowing machine.

"Where are you going to begin?" I asked after greetings had been exchanged.

"Well," he replied, "I was planning to start here at the top of the hill and work across that way," pointing eastward in the direction of the adjoining pasture. "Does it make any difference to you folks?"

"No," I said, "we won't be recording this morning so the noise won't bother us. I just wanted to give the field a last check before you cut it. I'll start at the bottom."

My point was that I wanted to listen for any new singers and catch them if possible. The situation was not critical because there were several other fields as yet unmowed, one below the Cannons' house and one adjacent to Professor Clark's house up the hill. Chances were that the fauna was similar in all.

Mr. Russell nodded, chuckled giddap to the horse, let down the cutting bar, and moved off to the rhythmic clatter of the blade. Aside from the humming of our own automobiles, this was the first mechanical sound we had heard in this secluded spot all summer. Thus spurred, I went to fetch my net, some collecting bottles, and my straw hat.

As I swished through the knee-deep grass, I noticed for the first time that the field shone more golden than green in the morning sun. The only sounds that I heard near at hand were those of ground crickets. From higher up in the field the only sound was that of the mowing machine.

The closest field lay just below the Cannons' house. I took a shortcut through a thicket of choke-cherries, hazelnuts, and briers separating the Pierce and Cannon properties. When I emerged into the sunshine, I sat down on a knoll to listen, first for general impressions, then for local details.

The first sound was the trilling of ground crickets; that was expected. They were close at hand, almost at my feet; they were loud; they were numerous; they were familiar. It was even possible to single out by ear enclaves of Striped Ground Crickets among the more numerous populations of Allard's Ground Crickets. The

Haying

field throbbed with the trilling and chirping. I closed my mind to this background music and searched with my ears for anything different. Sooner than I had expected, I picked out two other sounds, a very faint wheezing close at hand and a louder high-pitched buzz in more distant parts of the field.

Acting on the principle that one in the hand is worth two in the bush, I zoned in on the weak, little (I was assuming that such a small song must come from a small insect) singer. I got a line on where he was, memorized the apparent location, moved slowly and quietly to another spot, and got another line of sight—actually of sound. With this triangulation, I fixed my eyes on a spot and stared. I saw nothing, but of course I did not know what to expect. After nearly twenty minutes of stalking I finally discovered a small, exquisitely dainty grass-green and yellow grasshopper with long filamentous antennae and long tapering wings exceeding the length of his body. His wings were out of focus. I knew immediately that they were blurred because he was singing. It was almost incessant, but so faint. It was of a delicacy to match his appearance. When interrupted he began again with a slow tse-tse-tse, followed by a pse-e-e-e-e which often lasted as long as twenty seconds. He was the Slender Meadow Grasshopper, one of a group of three species to be found in New England.

Having spent weeks watching short-horned locusts and almost coming to appreciate their robust, sturdy, sensible, rustic appearance, I was unprepared for the delicate, elegant appearance of the Slender Meadow Grasshoppers. The difference in appearance reminded me of the difference between the sturdy working peasants in the stolid Dutch scenes of Avercamp and the frail dainty aristocrats with whom Watteau peopled his romantic landscapes—the real and the romantic.

I was getting a cramp from my rigid position, so I moved. The Slender Meadow Grasshopper stopped singing instantly and took to flight. But now I knew the identity of my quarry. Rising, I strode across the field sweeping at the height of the tall supple grass. Within a few minutes I had among the debris in the net about twelve meadow grasshoppers. Since I was not far from the laboratory, I choked off the end of the bag and transported the catch for later transfer to cages. I was eager to return to the field in search of the louder singers.

When I dumped the catch into a cage I discovered that I had specimens not only of the Slender Meadow Grasshopper but also of the Short-Winged Meadow Grasshopper. Aside from having wings that do not reach to the end of the body and being chunkier than his long-winged relative, this fellow is similar in appearance. Both species are grass-green, with a brown stripe bordered with white or yellow running down the back.

The songs of the two species are similar in structure and faintness. That of the long-winged species is a trill of pulses lasting two to five seconds and separated by two to twenty-six clicks, the whole sounding like tse-tse-tse-tse-tse-e-e-e-e. The song of the short-winged species is much the same but with only one or two clicks.

These were details that I learned later, because as soon as my net was emptied I returned to the field, where I headed toward the taller grass at the far end. There, loud singers were still active. The first song on which I zeroed in was a buzz, a zeeeeeee that continued for about three seconds, then stopped for about five seconds. Before commencing again, it was preceded in the interval by a dozen or two zips. This pattern of zips and buzzes continued without interruption for the five minutes that it took me to locate the singer.

I finally saw him perched about a foot down from the head of a stalk of grass. In appearance he was a large version of the Slender Meadow Grasshopper but proportionally more robust. His general color was green. He apparently saw me about the same time that I saw him, because he stopped singing, sidled around the stem, flattened against it, and straightened out his hind legs in line with the stem, thereby making himself less conspicu-

Short-Winged and Slender Meadow Grasshoppers

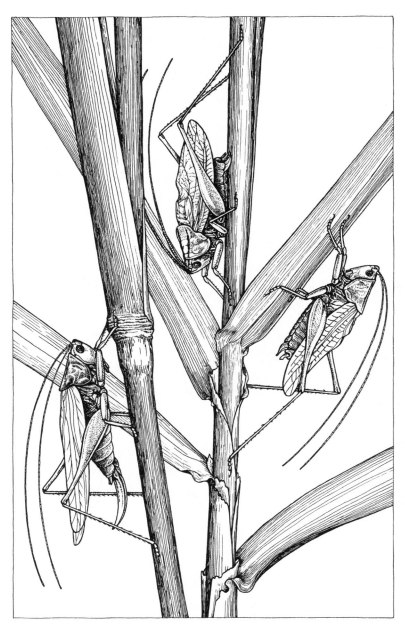

Common Meadow Grasshoppers

ous. On further inspection he turned out to be a Common Meadow Grasshopper, one of three species found in great numbers in New England. He is one of the most widespread of all meadow grasshoppers, and his song is one of the signs of midsummer. If any grasshopper can be said to be the mascot of the meadow, he is it. Even the scientific name of the genus to which he belongs, *Orchelimum*, is indicative of his character. In Greek it means "I dance in the meadows." And so he does, day and night.

A few days later all the fields had been mown, the fallen sheaves raked into windrows drying in the sun. Haying weather held, and within a week the hay was pitched into the wagon and the high sweet-scented loads were drawn away to the barn.

There lay the shorn fields, as exposed as a man who has just left the barbershop. The wavy locks were gone; the grass was now stubble. This was small discomfort to the ground crickets and locusts, but I wondered what would become of the meadow grasshoppers. Some would find refuge along the fringes of the fields where the mower had not reached, or in the low spots that the cutter could not trim. For the rest, there were no stalks from which to sing, no places in which to hide.

Mount Washington Interlude

Night's candles are burnt out, and jocund day
Stands tiptoe on the misty mountain tops.
Shakespeare, Romeo and Juliet, III.9

Three species of large meadow grasshoppers are found through-out New England. The Common Meadow Grasshopper, which we had already studied, lives everywhere east of the Mississippi River except in the extreme southeast; the Gladiator Meadow Grass-hopper populates a broad band of the northern United States stretching from coast to coast; the Dusky-Faced Meadow Grass-hopper occupies the coastline from Maine to Texas and a limited area south of the Great Lakes.

Despite intensive search we had not found the Gladiator in the vicinity of Franklin, nor were there any records of its having been taken in the surrounding areas. There was, however, an old published account of specimens having been captured on Mount Washington. On the strength of that report, we had decided to make a one-day excursion to hunt for him there. Unfortunately the exact site of capture had not been noted.

That summer, I made an expedition to Mount Washington along with a colleague, a visiting associate of Professor Pierce. We planned our departure from Franklin so that we would arrive in Pinkham Notch about dawn and thus have an entire day to climb and search. The object of our quest was the Gladiator Meadow Grasshopper.

Our plan was to climb the mountain by way of the New Fire Trail, into the basin of Tuckerman Ravine, and thence up the headwall to the summit. Most of the climb would be in the forest, so we would be able to enjoy that part of the ascent without having to be preoccupied with professional duties. Not until the timberline would there be open spaces congenial to meadow grasshoppers.

I had last made the climb to Tuckerman Ravine the previous spring. Skiing there was traditionally good in April; it was an ideal time for facing the challenge of a schuss down the headwall, a slope of about fifty degrees. At the time of my skiing trip, there had been an overnight blizzard. The depth and lightness of the new snow had made climbing difficult. A small band of us had struggled upward on skiis with sealskins attached; we had found the going slow and exhausting.

Obviously the ambience of the summer climb was different, not solely because of the absence of snow and freezing temperatures but because of the scented summer air, the sounds of summer, and the subtle seasonal differences in the clouds and sunlight. Together, all these whispered summer.

Our climb had begun in mixed hardwood and conifer forests, in still air spiced with blended scents of spruce and fir. As we progressed higher we passed through denser stands of evergreens, where the perfume of the forest became more insistently perceptible. We paused occasionally to catch our breath and to absorb with eyes and mind the dark beauty of the forest.

At one rest stop we heard someone approaching from down trail. In the immense stillness, sounds, even of footfalls, travel far. Minutes passed before the climber, a gray-haired man, came into view. He was back-packing a crate of oranges, presumably to the Tuckerman Ravine Shelter. Supplies for the summit would have been delivered either up the automobile road or on the cog railway from Crawford Notch.

He acknowledged our greeting without breaking his steady, measured pace. As he passed from sight, we resumed our climb. At one point we were overtaken by a young couple climbing at high speed. Later we passed them as they rested. This sequence of mutual passing repeated itself several times over the next two hours. We never caught up to the gray-haired man.

We left behind the sounds of the lowland forest as we advanced higher. At low altitudes chickadees had been chattering and red squirrels scolding, but now we were in a zone of silence and heat. At ground level in the shade, where the trail snaked its way among the ledges, there was no air movement to provide relief from the heat. We were grateful for the shade.

As we reached higher altitudes, the silence was ruffled by the murmuring of the forest. A cool wind flowing down from the mountain ridges was stirred into small vortices by each twig and needle that it encountered. These vortices produced faint sounds, but the combination of hundreds of these aeolian noises caused the whole forest to whisper. Each tree had a different voice, and a voice for each season. The bare hardwood of winter presents to the wind thousands of twigs of many sizes. These arboreal harps sing with a multitude of low tones. The needles of conifers, being more uniform in size, cause a limited range of high tones when played upon by the wind. The twigged maple moans while the needled pine whispers. Together the voices of legions of trees join in concert, such that the whole forest murmurs.

That day the forest murmured gently as we pushed up toward the timberline. It seemed as if the gentle, vulnerable life around us were rebuking us for our intrusion. There were no animal sounds other than ours, but as the trees whispered I could understand how the people of earlier ages were awed, intimidated, even frightened by the forests, not alone by their labyrinthine mysteries, their darkness, their fancied inhabitants, but by the forest murmur. Small wonder that early peoples, knowing forests to be alive but immobile, anthropomorphized. This was their kind of forest through which we passed, an enchanted botanical world which whispered and moaned.

At eleven o'clock we broke out of the enchantment at Hermit

Tuckerman Ravine in July

Lake. We had reached the elevation where the forest, thrown like a green robe over the shoulders of the mountain, had come to a frayed end. Here it thinned out. Here the trees were shorter, more gnarled, more misshapen as a result of the constant strong winds.

The view of the ravine was spectacular. Advanced though the summer was, a massive snowfield clung to the headwall, white, clean, and sparkling. Ninety minutes later, however, when we reached it, we saw that it was old, gray, and grainy from repeated thawing and refreezing. This "corn snow" signaled that winter was in its dotage.

Close to the pond there was a small lean-to shelter, its open side in the lee of a large boulder. Nearby trickled a meltwater stream, from which we drank. In those days the mountain was less frequently visited than now, and pollution was nonexistent. We decided then and there to break for lunch.

We had traveled light; in our packs were some sandwiches, two cans of pineapple juice, a chunk of cheddar cheese, some chocolate bars, small boxes of raisins, and, of course, a camera, a collecting net, and specimen bottles. Around the pond the grass had found enough sun to encourage growth in small swards. We chose one of these spots for lunch, happy at this cooler altitude to exchange shade for sun.

During lunch I attended to business—that is, I listened for insects—but heard nothing; nor when I later swept the grass did I find any orthopterans in the net. Apparently the spot was not a suitable habitat for grasshoppers. We therefore resumed our climb, up the little headwall, into the bowl of the ravine. Here the stands of dwarf trees, which had been covered with many feet of snow during the winter, presented an almost impenetrable knee-deep to shoulder-high thicket. Matted dwarf spruce, birch, and willow made progress virtually impossible whenever we strayed from the marked trail.

I heard no singing whatsoever, but I did notice a sluggish olivaceous grasshopper crawling on one of the boulders that stuck up above the vegetation. He was moving so slowly that I was able to pick him up between thumb and forefinger. I suspected that he might have been more agile had the day been warmer. I recognized him as the famous White Mountain Wingless Locust, discovered and named by Scudder in 1863.

The history of this grasshopper, along with that of the even more famous White Mountain Butterfly, is fascinating. The White Mountain Butterfly in the eastern United States lives only on the tops of the higher mountains—Mount Washington and neighboring high peaks in New Hampshire, and Mount Katahdin in Maine. Farther north it has a worldwide distribution around the North Pole. The grasshopper is restricted to mountaintops in the warmer parts of its range but in more northerly areas inhabits lower levels, where there is boreal vegetation. It is a lover of thickets, especially those of dwarf birch and blueberry.

The peculiar distribution of these two insects is a consequence of glaciation. As the ice sheet advanced southward in the last Ice Age, animals fled south ahead of it. As it eventually retreated during thousands of years of warmer temperature, animals that were accustomed to life in northern climes fled the advancing warmth behind them and followed the coolness north across plains, along river valleys, over hills, and up mountains. As warm climate continued to follow the ice, insects that had climbed mountains were forever trapped by a vast surrounding sea of heat. Those that had not escaped to mountains either died in the heat and the encroaching southern forests or continued to move northward to their ancestral homelands.

The glacier grasshopper was interesting because of his history and his winglessness, but without wings he could not sing. From the point of view of our study, he had no secrets to reveal. I returned him to his rock.

We contemplated continuing to the summit, where we knew there were alpine meadows, but decided first to explore more carefully parts of the basin which even the trees found inhospitable.

Given the lateness of the day, we were not enthusiastic about climbing higher. I still heard no singing but decided nevertheless to sweep some of the low vegetation and especially those places where the grass was tall.

To my great delight I netted two meadow grasshoppers, one of each sex. The male looked very much like the Common Meadow Grasshopper, except that his wings extended farther beyond his abdomen. This difference turned out later to be a matter of individual rather than species variation. The Common Meadow Grasshopper and the Gladiator are indistinguishable except for some very minor differences in spines. The most striking difference was revealed to us later: it was in the song. Whereas the song of the Common Meadow Grasshopper consists of sequences of trains of pulses separated by clicks, the song of the Gladiator is made up of trains of pulses without intervening clicks. Both songs are at the same high frequency.

Two grasshoppers, looking alike, singing songs that differ only by the presence or absence of twelve clicks between trills! It seems to be a trivial reason for climbing a mountain, but we did not know in advance that the difference was so minor. To human beings it did indeed seem trivial; for the survival of the species it was not.

The song serves to bring the sexes together. Imagine being on a mountaintop or in a field of grass, and being small, less than an inch in length. Imagine the formidable difficulty of trying to find a mate in those vast uncharted spaces. All the singing insects face this challenge.

Calling-songs are the solution to the problem. Appearances are neither distinctive nor meaningful, but an acoustic signal to which the female can orient and home is essential. And the song must be correct in order to lead the female to the genetically compatible mate, otherwise consummation, even if completed, is fruitless. The small musical differences may be inconsequential to us, but to the females they are crucial.

Our excursion was not an exercise in triviality or futility because it taught us a bit more about how species are protected against miscegenation and how the rich diversity of the natural world is protected. The venture was, from a selfish point of view, an enrichment of our perception and appreciation of nature.

Our descent to Pinkham Notch took about two hours, as we hastened to outrun the lengthening shadows. My thoughts detached themselves from the moment; the experience would be savored in memory. I thought back to my skiing trip in April, when my friends and I had likewise raced darkness to the bottom. That descent over the snow had been measured in minutes.

The Salt Marshes of Cape Cod

In summer-days, like grasshoppers rejoice,
A bloodless race, that sound a feeble voice.
Homer, Iliad, III.199, tr. Alexander Pope

The singing of our new acquaintance from the mountain had been recorded and analyzed, his file examined under the microscope, his behavior observed and photographed. While he sang unabashedly in his private quarters indoors, his close relatives, the Common Meadow Grasshoppers, together with Small Meadow Grasshoppers and ground crickets, were in full voice in the fields. They would continue until the first mass of cold air from the north stilled their exuberance. The grasshopper in the warmth of the laboratory would persist until aging muscles and diminishing nervous energy slowed him to final silence.

From one point of view, we did him a favor by keeping him in protective custody. His life expectancy in the wild would have been short. There were many mouths to feed—mice, snakes, birds, carnivorous beetles, spiders, and ants. There were fungal and viral infections. There were mists of pesticides blowing from cornfields and orchards. On the other hand, our care could not protect him from his natural mortality. To each animal there is an allotted span, and there is no escaping one's temporal limits. Large animals tend to outlive small ones, and those animals most prodigal with their expenditure of energy have shorter lives than those which metabolize frugally with slow efficiency. The singing Orthoptera are both small and energetic. The summer of singing, which represents less than one-half of one percent of a human lifetime, is two-thirds of the lifetime of a grasshopper. Perhaps Aesop's ant was wrong when she chided the grasshopper for spending his life singing. What other future was there for him?

I listened to the field but detected no unfamiliar songs. Had I missed anyone? We had explored the grasslands and the bogs, the waste places, ledges, and gravel pits. The forests had thus far been silent. We had sampled alpine meadows. I knew that as the summer advanced and fall approached, other species of singing Orthoptera would attain maturity, but at the moment this was only the season's promise.

In the meantime our thoughts turned farther afield. Against all odds, the short excursion to Mount Washington had been successful. With that in mind, we began to think about other habitats not found in southern New Hampshire. In particular we wondered about salt marshes.

New Hampshire does have some coastal marshes, but we reasoned that more southerly locations might enhance our opportunities to catch species for which New Hampshire was a northern limit of range. Furthermore, memories of broader reaches of marshland and association with sand dunes drew my thoughts toward Cape Cod. Accordingly, with Professor Pierce's blessing, I set out for the Cape. My destination was Little Buttermilk Bay.

As a child I had spent many summers on Cape Cod, first in Dennis on the Massachusetts Bay shore, later on the south shore at Buzzards Bay. My sojourns there, together with many automobile rides throughout the Cape, had given me a sense of the unique and lovely nature of this land of sandy beaches, mud flats, dunes, bogs, and dry pine and oak forests.

Part of the figurative atmosphere of the Cape is its actual atmosphere. The scrub oak and pitch pine forests effuse a quite distinctive scent, different from that of white pine forests and of the spruce and balsam fir forests of New Hampshire. Bushes of the bayberry thickets do not have to be crushed to release the aroma so characteristic of the Cape in fact and fancy. The wild beaches on the Massachusetts Bay side give off, as might be expected, the smell of the open ocean, but it is delicately different from the open-ocean smells of Maine's rocky coast and of the tropics. The sand itself on hot days releases all the locked-in effluvias of half-buried skate and whelk egg cases, of stranded angler fishes and dogfish, and of shells in the process of being bleached clean. Behind the dunes and on the Buzzards Bay side, there is the unforgettable smell of the flats. This was the dominating smell of Little Buttermilk Bay, but, I hasten to add, the better smell of flats. To some people the term "mud flats" conjures up visions of decay, dead things, spoilage, and uncleanliness. To me the odor of mud flats conjures up visions not only of heaping dishes of fried clams, but of salt-marsh grass flecked with colored streaks of pollen, of Great Blue Herons statuesque as they stand patiently waiting for small fish to leave the safety of the eel grass, and of sea fogs and all things littoral.

Little Buttermilk Bay is an arm of Buzzards Bay and is connected to it by a narrow inlet constricted by the abutments and piers of a railroad bridge and a highway bridge. The tide rushes through this channel with a force and speed that challenges even the hardiest rower. The bay itself is shallow. At low tide it is possible to wade around in a central sand and mud flat where quahogs once abounded. The shores are low, reached from the seaward side only through gardens of eel grass. At low tide there are exposed mud flats of the softest, stickiest, thickest kind. Then, still approaching from the seaward side, one encounters a fringe of tall coarse *Spartina* grass. The common name is cord grass because a Mediterranean species was used to make cord. As the marsh rises from the water's edge, this tall grass is supplanted by a very fine, short *Spartina* grass which characterizes the broadness of the salt marshes. On the land side these merge into dunes, and the dunes in turn blend with beach plum, bayberry, and eventually dry scrub oak and pitch pine.

On my first day I set out to explore by ear and by sweep-net

Cape Cod salt marsh

the *Spartina* salt marsh. During the warm sunny forenoon I heard nothing of particular interest. A light sea breeze stirred the *Spartina*, and the sea of grass waved in emulation of the sea waves, but it was a hushed sea.

At noon I wandered onto a small sand spit to have lunch and to relax in the beauty of the day. Fiddler crabs running in all directions near the waterline and among the stalks of grasses in the wet mud flat were semaphoring madly to one another. Some Laughing Gulls, wary of me, were scavenging on another part of the beach. Terns were fishing offshore. I was seriously contemplating a swim when a shift in the breeze carried to me a soft hushing sound quite distinct from the more silken swish of the grass.

At first I thought the sound was the singing of Small Meadow Grasshoppers, either the Slender or the Short-Winged Small Meadow Grasshopper. I dismissed the sound from mind because we could obtain those species in Franklin. But then I had second thoughts. I was not down here on vacation; my conscience said I should investigate. The question was: Should I stalk the singers, or should I sweep the grass in great swaths?

I tried the stalking stratagem first, and after about fifteen minutes of advancing one step at a time, stopping, looking, listening, I finally spotted a grasshopper singing softly, unperturbed by the swaying of the grass stem to which he was clinging. The sighting was a disappointment; the grasshopper appeared to be the short-winged species which we had already studied; nevertheless, I swept him from his perch, together with several others I had not noticed. As a matter of course, I transferred several to a jar; one I picked out for closer examination. My meticulousness was rewarded; he did not look quite like the grasshoppers from Franklin.

I have mentioned before that the amateur is frustrated by many of the keys for identification constructed by professional entomologists. The keys often depend on minutiae. I have also warned against an overdependence on coloration, because of the wide variation among individuals of the same species, not only from one geographic range to another but also within a given local population. Some species characteristically have both a green and a brown form. This is not to say, however, that one should be insensitive to coloration. Here was a case in point.

The predominant color of the two Small Meadow Grasshoppers from New Hampshire was grass-green. The newly captured specimen was short-winged, which alone would have marked him as the Short-Winged Meadow Grasshopper, *but* his color was vivid, a faintly bluish green. When I checked back in my reference books, I determined that he was indeed a new (to us) species, the Salt-Marsh Meadow Grasshopper. He and his fellows were destined to accompany me back to Franklin. There we would study his song. At the moment he sounded very much like his relatives. This experience reinforced a lesson learned early in my career as a biologist: never take anything for granted.

The following day, late in the afternoon I tracked down two new grasshoppers, one in the field next to the cottage where I was staying, the other in the salt marsh itself. The time was half-past three in the afternoon. I had found nothing of interest in the marsh, and so had come up from the shore to listen in the woods adjoining the house. I never got to the woods, because on the way I heard some buzzing in tall grass at the edge of a moist ditch.

The song was soft and lisping, somewhat like that of the Common Meadow Grasshopper and the Gladiator. It continued uninterruptedly for minutes at a time, a soft tsip-tsip-tsip or ik-ik-ik-ik, depending on one's personal phonetic interpretation.

The singer was not difficult to find. He was perched head down near the top of a sturdy stem of grass. He did not startle easily, but when he judged that I had come too close, he plunged head first down the stem to the ground, where he landed on his head and remained motionless. When I picked him up, I knew immediately what he was: a conehead, and a particular conehead—the Sword Bearer.

The coneheads, as the name suggests, are second to none so far as bizarre appearance goes. Their faces are acutely slanted chinward, as are those of many meadow grasshoppers and locusts. But accentuating the slant of their face is the startling conical projection of the top of the head. It is reminiscent of the shapes of heads of fictitious aliens as commonly depicted on television. There is no mistaking a conehead, whether he be an imagined alien from outer space or a very real earthly grasshopper.

So much for his family name; his species name, the sword bearer, is derived from the appearance of the female. The time has come, therefore, to speak of ovipositors. Throughout this book I have used the masculine pronoun when referring to individual orthopterans. This is not blatant male chauvinism. The use is simply a matter of accuracy because, except for some female katydids that emit various squawks, only male orthopterans sing. When it comes to matters of identification, however, the females of the species come into their own. The most distinctive diagnostic characteristic of Orthoptera is the ovipositor, the tube for the passage and positioning of eggs in the soil, in grass stems, into rotting wood, or within living bark. Each ovipositor is designed for the unique egg-laying requirements of the species.

This neat accommodation between structure and function is vitally important for the propagation of the grasshopper and most convenient for entomologists. It also frequently poses dilemmas. If two species are separated on the basis of the shape or length of the ovipositors, how is a person to identify males, especially if the males of different species are nearly identical in appearance? As we have learned so far, songs are often better diagnostic characteristics than structure. But only males sing and only females possess ovipositors. Who is related to whom? The solution to this dilemma we must leave with the specialists and hope that they are as astute on this score as are the insects themselves. Once you see the female Sword Bearer, you will agree that she is well named.

The other conehead I found that night in the salt marsh was

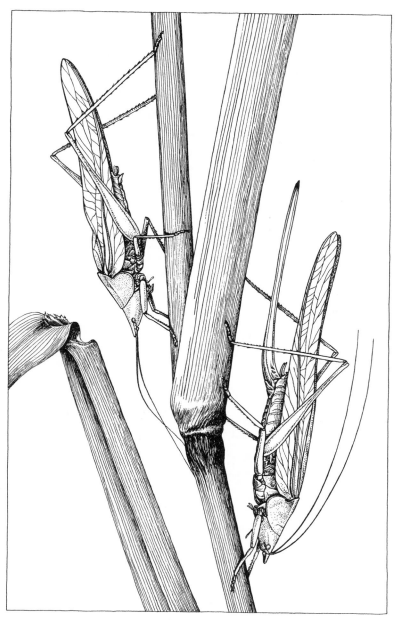

Sword-Bearer Coneheads

also easily identified. Just after dusk had fallen and the evening
star had appeared in the blue-black sky, there arose in the marsh
a clatter as of the singing of swarms of cicadas. Obviously no cicadas, sun-loving "hot bugs" that they are, would be singing at this
late hour. There was only one orthopteran that could rival the
cicada, the Robust Conehead; so off I went, armed with net,
bottles, and flashlight. I forgot one thing: insect repellent.

The night was beautiful. There was little chance of arousing
the suspicions of people in beach cottages, because mine was not
the only moving light. Eel fishermen with lighted jacklamps in the
bows of their skiffs were gigging for eels, and my light could easily
be mistaken for one of those. It was also possible that more imag-
inative souls would think that ignis fatuus, the will-o'-the-wisp of
low swampy places, was abroad in the marsh that night. The haz-
ard I faced was not the local constable but the chance of stumbling
into a muddy tidal pool or being eaten alive by the no-see-ums.
As on all warm nights, these minute midges were ferocious.

The conehead was neither difficult to locate nor wary of my
approach; the light did not bother him a bit. He was enormous
by grasshopper standards. I transfixed him with my jacklight. He
kept singing. I moved closer to observe him in action, stopped to
scratch some no-see-um bites, and inadvertently jarred the clump
of *Spartina* that was sheltering him. The conehead stopped, hesi-
tated, and, like the smaller Sword Bearer, dropped head first to the
ground. He presented a ludicrous picture, almost as though he had
pegged his head into the ground. He did not resist when I picked
him up by the wings. I was amused by the thought that this loud
boaster seemed to lack the spirit of the smaller, gentler inhabitants
of the marsh.

My visit to the Cape had yielded three species to study.
Three days had been allotted for my excursion, and I knew that
there was another species to be had, even though I had not heard
him around Little Buttermilk Bay. I decided, therefore, to go the
following morning to the salt marshes and dunes at Sandwich.

When I arrived, late in the afternoon because I knew that the singer I sought was a late riser, I settled down at the edge of a marsh behind the dunes. The beach itself was empty; the sun hung on the horizon and most of the bathers had left. The no-see-ums would soon be out in force. Putting that thought out of my mind, I began to listen. I could hear the surf on the other side of the dune, a Laughing Gull or two, and the squawks of some passing Night Herons that had left their rookery early. The salt marsh itself was quiet.

The sun finally dropped from view behind the landward forest. Only weak rays filtered through the tops of the pines. Shadow crept across the marsh. At that moment the marsh began to come to life. Fortunately the first sound was not the Robust Conehead, because his song would have drowned out softer ones. I could tell there were no Salt-Marsh Grasshoppers close by, because I would have heard them; the more distant ones were inaudible.

In the quiet of dusk I heard a few clicks, as though a musician were tuning his instrument. Then came long trains of pulses, tzip-tzip-tzip, very rapid, with many clicks between trains. I swept the tall cord grass and had my singer. Although he was a close relative of the Common Meadow Grasshopper and the Gladiator, his appearance was distinctive: slender, dun-colored, and gray-faced. Hence his name, the Dusky-Faced Meadow Grasshopper. With that sweep I had caught the fourth of the meadow grasshoppers and coneheads that I had set out to find. I also had an impressive collection of no-see-um bites.

With the exception of the din of the Robust Conehead, the songs of the salt-marsh grasshoppers are soft and gentle. In the aggregate they provide a delicate lisping background to the more musical trills and tinkles of the crickets. The effect is akin to that of people humming along with musical instruments. In the salt marshes in particular, where cricket populations are small, the grasshoppers set a different mood and are as much a characteristic part of the marshes as are the smells, the sea of grass, the salty mud, tidal currents, the high humidity and temperature, and the broad openness bounded by sea and sky. What it is about this land that renders it so felicitous to some species of grasshoppers and unattractive to close relatives which are almost indistinguishable in form, color, and behavior remains one of the least appreciated wonders of nature.

A Counterpoint of Scrapes and Squawks

The seasons change, the winds they shift and veer;
The grass of yesteryear
Is dead; the birds depart, the groves decay:
Empires dissolve and peoples disappear:
Song passes not away.
William Watson, "Lachrymae Musarum"

Summer was in its prime, but already portents of approaching autumn could be felt in the air, the wind, and the appearance of the sky. Cloud patterns were different; the arc of the sun's path bent lower to the horizon; the spectrum of light seemed more yellow. New colors reflected from the landscape. The green world was less vividly green; it had assumed a deeper hue. Hints of yellow and brown burnished the leaves; the colors of exuberant growth turned sombre. Everywhere there was change.

Cycles of reproduction had completed an annual revolution, had reached an appointed stage at an appointed time. In the plant world, hay and early crops, peas, beans, lettuce, were being harvested. Grasses were heavy with seed. Blueberries added color to the barrens, and raspberries brightened the cane thickets. In the animal world, eggs had hatched, fledglings had flown, young had been born and weaned. In the world of orthopterans some, like the Northern Spring Field Crickets, had sung out their lives. Others, their eggs stored for the winter, were approaching their autumn of life. The songs of aged males had lost their allure. With the decline and passing of these species there would come others which, marking time by a different calendar, would only now be tuning their instruments for August concerts. Among the earliest would be the bush katydids and their family relatives the round-headed katydids.

Their songs would add an awkward counterpoint of scrapes and squawks to the lulling strains that had characterized the rare days of June. Theirs would be a less joyous and less soothing music. Yet, when it did come, it seemed to suit the impending change of season.

When we had commenced our studies in early June, the fields had throbbed with the clear music of chirps, melodious thrills, and bell-like tinkles. The bare, rocky, arid places were left, as befit their nature, with a barren music of shuffles, flutterings, and crackles—nature's attempt at minimalist music. The thickets and woods were silent except for the trills of the Sphagnum Ground Crickets in the bogs and the Tinkling Ground Crickets in the leaf litter of the forest floor.

As the clear hot days of July had incubated the land and as the grass had grown longer, variations had been added to the music. The meadow grasshoppers strummed a background of swishes, buzzes, and whirrs, softly modulated, more or less continuous, and punctuated with delicate staccato clicks. The mood was light, carefree, even joyous.

When I returned from the Cape after only a week's absence, I was immediately struck by the changes. I was greeted by a ragged asynchronous medley of rough, grating sounds emerging from the bushes—huckleberry, choke-cherry, bayberry, hazel, alder, wild rose—and from clumps of tall coarse grasses and weeds. To my ears the sounds were uninspiring, but obviously they were not so to the katydids. After all, the scrapes and squawks had been effective in attracting females for millions of generations. Music is as much in the ear of the listener as beauty in the eye of the beholder.

My first encounter with a bush katydid came late one afternoon. I was wandering (most of our official work at Franklin consisted of wandering) through tall goldenrod on a west-facing slope where the lighting was especially bright. Ahead of me I heard a rather loud series of sh-sh-sh-sh-sh pulses. There was no regular pattern to this call; sometimes there were two-pulse trains; other times, fourteen. From other parts of the slope I heard similar calls, timed as though neighbors were calling to one another. Although one of the services that the calls perform is that of alerting females to a caller's presence, the calls might also keep males comfortably spaced in their habitat. The males do answer one another. They will also answer an imitation sound made by a human being. The female's answer is only a discreet click.

I approached the nearby katydid too carelessly and must have blundered uncomfortably close, because he took to flight. He sailed silently, rather slowly, somewhat butterfly-like, in a long zigzag traverse of the slope. At the far end of the traverse he plumped into an alder.

Rather than pursue him I sought another individual nearer at hand. This time I acted more cautiously and saw him before he saw me, or at least before he was alarmed by my presence. I could appreciate him in his natural pose. He was the first katydid of any kind that I had seen. His size surprised me. He was nearly two inches from head to wing tip, ungainly, with a disproportionately small head and disproportionately spindly hind legs. Their length was impressive and not at all commensurate with his feeble jumping prowess. They seemed more of an encumbrance for a creature that spent his life clambering about in thickets.

His pea-green wing covers had a shape to match their color: they resembled flat pea pods. Nor were they much more active than pea pods. They are indeed the katydid's musical instrument, but they are essentially useless in flight. Aside from producing sounds, they have enormous value as camouflage. Their podlike or leaflike appearance provides protection from predators only so long as their owner remains still and, to mix a metaphor, keeps his mouth shut. Perhaps that is why nature has arranged that he confine most of his calling to nighttime.

I was able to catch this "beauty" between thumb and forefinger. His song, sh-sh-sh-sh, his narrow wings, his pinched face, and

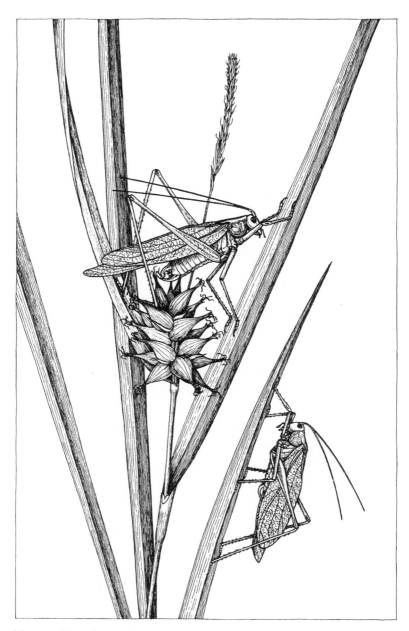

Texan and Broad-Winged Bush Katydids

his "tail plate" marked him as a Texan Bush Katydid. The characteristics of this katydid are distinctive; his name is misleading. His species extends all the way from New England to Wyoming and south to Florida and western Texas. The one I held was, of course, a born and bred New Englander.

When it comes to identifying katydids that live in bushes, a collector is provided with three clues—namely, the song, the shape of the head, and the shape of the tail. The variety of head shapes would titillate a phrenologist if he chose to philosophize about orthopterans. I doubt whether an examination of all the different heads available would reveal anything about the mental faculties of orthopterans, but it would surely identify the species of which the owner was a member.

In the realm of heads, there are the rotund heads of the field crickets and ground crickets, the slanted, long, pointed, pinched heads of the meadow grasshoppers, the surrealistic heads of the coneheads, the solid aquiline heads of the locusts, and the two distinctive kinds of heads of the katydids that live in bushes. The bush katydids are born with heads that are narrow between the antennae; the Round-Headed Katydids, as the name implies, have round heads. Nobody seriously believes that these geometric features have anything to do with cerebral excellence or lack thereof. As a matter of fact, most of the head is filled with muscle to work the jaws, perhaps indicating that eating is more important than thinking. Nevertheless, heads are reliable aids to identification.

Once a katydid has been identified as narrow-headed or round-headed, the next task of the collector is to determine which species has been captured. Here one turns to tails. Among crickets and grasshoppers, only females possess posterior distinguishing features prominent enough to be recognizable without a hand lens. The Sword Bearer is a striking example. Bush katydids also possess distinctive ovipositors. The males, not to be outdone, have projections of their own. The male of the species can be identified at a glance by the shape of the "tail." This elongation of the last segment of the abdomen may be decurved, broadly forked, or narrowly forked, with a central spine or without one—each species has its own configuration. What the functional significance of these structural idiosyncrasies may be is anybody's guess.

Armed with information from Blatchley about heads, tails, and habitats, together with assurances that five species of narrow-headed katydids and two species of round-heads had been taken in New Hampshire at one time or another, I set out to discover whether or not they could be found in the region around Franklin. Most of my field work on those August days was shifting from early afternoon hours to very late afternoons, when the shadows were long and twilight came early. Later in the month I would concentrate on nighttime collecting, because many of the orthopterans of late August and September specialized in nocturnes, but now late afternoons suited the timetables of most.

On this revised schedule, the first new katydid to fall into my grasp was one of the most common New England species, the Curve-Tailed Bush Katydid. He is also the most robust of the katydids and is as long as the Texan. He is easily distinguished from the latter by the absence of a spine in the notch of his tail. Another characteristic feature is the breadth of his wings.

I was able to run him down in some alders bordering a swale below the laboratory by zoning in on his loud, simple call. Various authorities have described it as a forceful, brief, and infrequent zzikk or bziwi. To any ear it is a scratchy, squeaky note, the sort of noise made by scraping a thumbnail over the teeth of a pocket comb. Since the noise is made only at long intervals, considerable patience is required to locate the creature. The rest is easy. Like all bush katydids, he moves slowly. When I picked him up between my fingers, as I had the Texan, he did not even struggle.

Some thirty years earlier, Scudder had observed that this species had different daytime and nighttime calls. The former is given only in bright sunlight and changes to the nighttime song whenever a cloud passes over the spot. That song is more repetitive—slow trains of eight or more pulses.

Oblong-Winged and Round-Winged Round-Headed Katydids

Another common katydid that tried my patience was the smallest of the five species, the Fork-Tailed, so named because the male's "tail" looks remarkably like a miniature cloven hoof. I found this fellow in tall coarse grasses and huckleberry bushes, whence his sharp two- or four-syllable zeep rang out above the background noises at long irregular intervals.

Several days passed before I heard any new calls from the bushes, even though I explored wide and far from the laboratory. The fields were still vibrant with trills and chirps, a fine music consonant with a habitat of fine, resilient, graceful grasses. In contrast, the inhabitants of the thickets emitted calls that were, in their musical form, angular, abrupt, and coarse, as were the physical forms of the thickets of twigs. Probably, lazy slow-moving creatures have a better chance of survival in twiggy jungles than in combed grasses.

In any case I found my next species in thick bushes bordering and even invading a field-stone wall. The late afternoon sky was so heavily overcast at the time that some of the nocturnal orthopterans had already begun their serenades. The Broad-Winged Bush Katydid that I finally tracked down is a northern species, broader in the wing than his congeners. The prominence of the wing veins makes his appearance even more leaflike than that of any other species. He is primarily nocturnal, but the waning afternoon had coaxed him into activity. His performance consisted of a toneless zikk-zikk-zikk of four to eight pulses given at irregular intervals. In the nighttime the call consists of longer series of trains made up of as many as eleven pulses. It is the loudest nocturnal woodland call at this time of year.

Thus far we had studied four of the five resident bush katydids. There had been no sign of the Northern Bush Katydid. I was unfamiliar with his song, and the books did not mention it. At the risk of being arrested as a suspicious character, I probed around thickets at night, always taking care to avoid populated areas. Stumbling around in the dark looking I knew not where for I knew not what proved to be a fruitless endeavor. I ran down every unfamiliar song that fell upon my ear. None came from him.

Two of the unfamiliar songs led me to the two endemic species of round-headed katydids. One, a rasping squawk with a terminal itzic repeated at intervals, came from the Oblong-Winged species, a frequenter of shrubs, small trees, and occasionally goldenrod. The other, a fast, soft, lisping tsip-i-tsip-i-tsip-i or chic-a-chee, chic-a-chee, chic-a-chee, was the call of the Round-Winged species, a denizen of low-growing shrubs, tall weeds, and grass.

Where was the Northern Bush Katydid? What was his habitat? Where did he live? My failure to find him did not mean that he was not around. I was reminded of some words of Edmund Burke: "Because half a dozen grasshoppers under a fern make the field ring with their importunate chink—pray do not imagine that those who make the noise are the only inhabitants of the field." Maybe the time had come to give up net-sweeping and listening and devise some other strategy.

The Rare One

My cottage lamp shines white and clear.
Joyce Kilmer, "The Twelve Forty-Five"

*I*n midsummer I gave up my lodging in town. A room in the center of Franklin and next to the railroad station was convenient for the limited diversions that the town offered, but it strained my pocketbook, and three meals every day at the Franklin Café palled. There were limited gustatory alternatives, whether at the café or at other eating establishments. Furthermore, the town offered few opportunities from an occupational and recreational point of view. Movies and bowling exhausted the latter. Insofar as the former were concerned, there were limits on the extent to which nocturnal entomological forays could be pursued profitably and with impunity. So long as I had no alternatives, living in town was tolerable; but when the Cannons offered me, gratis, the use of the professor's tiny studio, I accepted gratefully and with alacrity. The move not only enhanced my comfort, social life, and financial status, but it was directly in one case and indirectly in others instrumental in adding three more members to our roster of singing orthopterans.

The studio was a small shingled cabin with casement windows, a field-stone bench along one side, a very convenient rain barrel at one corner, and rustic comfort within. It was situated in pine woods two hundred feet or so downhill from the main house. On the downhill side it looked out on one of Mr. Russell's hayfields. One other feature merits special mention because of the role it played in subsequent entomological events. I refer to the carriage lamp outside the door.

I did not sever my connections with the town completely. There were errands to be run, visits to Surowiec's Market to buy forage for our insects and now for myself, weekend recreation, and swimming in Webster Lake. On the other hand, residence in the cabin allowed me the luxury of listening to a radio at night, of working in the laboratory at odd hours, and of access to a collection of books belonging to the Pierces. More to the point, it gave wider opportunities to explore the locality after dark in search of new songs. No longer was I restricted to the impoverished musical fare of the town—the sounds of the fast freight and of the Carolina Ground Crickets in the graveyard.

By night I searched the hedgerows, the woods, and the roadsides where there were no houses for any sound that might be that of the Northern Bush Katydid. Could it be that he did not live in the area? I doubted it. All his close relatives resided here. Did he frequent a different, perhaps bizarre habitat? That seemed unlikely. I had explored arid areas, moist areas, fields, bushes, and clumps of various kinds of weeds. Perhaps he did not sing. That, too, seemed unlikely. If he did not sing, how could his mate find him? Surely her task would be more formidable than mine, considering the darkness of the nighttimes and the vastness, from her point of view, of the habitat. I concluded that, as the authorities had stated, he was rare or at least elusive; and that was that.

While I continued to search and listen, I pondered the phenomenon of rarity. When one stops to think about it, one discovers, with some surprise, that most animal species are scarce in relation to the extent of suitable places for habitation and that many suitable places are empty. This is the paradox of scarcity and plenty. It applies not only to space but to available food. There are many nutritious plants available, but only a few are chosen by herbivores, even by those that are not epicures. They would undoubtedly have starved at a Lucullan banquet.

But given the relative scarcity of species in general, there are
still extreme degrees of rarity. This observation raises many questions. Why are some species more rare than others? Why do some
populations rise and fall cyclically? Are the rare ones on a one-
way road to extinction? With respect to the last question, Darwin
wrote in *The Origin of Species:* "To admit that species generally become rare before they become extinct, to feel no surprise at the
rarity of a species, and yet to marvel greatly when the species
ceases to exist, is much the same as to admit that sickness in an
individual is the forerunner of death—to feel no surprise at sickness, but, when the sick man dies, to wonder and to suspect that
he died of some deed of violence." Why the Northern Bush Katydid was rare and whether extinction lay ahead of him were nag-
ging questions. Perhaps he was one of nature's less successful experiments. On the other hand, perhaps he was unusually
successful in that he could survive as a rugged individualist.

Generally speaking, the abundance of a species in any area
reflects the outcome of interactions between the inherited characteristics of the species and influences of the habitat it occupies.
Acting together, these forces determine the birth rate and the
death rate, and hence the abundance or rarity of individuals. Birth
rates are determined by fecundity—that is, the maximum number
of eggs a female can produce. Fertility is the actual number produced. In the best of all worlds, a female may be able to produce,
for example, five hundred eggs. In the real world this number
might be reduced by limited available food, inimical weather, or
other environmental events to one-third that number. Death rates
are also determined by the interplay of environmental factors, disease, predation, parasitism, and the inherited ability of a species
to cope with them. Different species may have different degrees
of susceptibility and immunity to disease and different degrees of
aggression, agility, or skill in fending off predators.

Which of these interactions and in what proportions are responsible for the rarity of the Northern Bush Katydids is not

known. And even if I had known, that knowledge would not have gone far toward enabling me to ferret out the creatures. I continued to search.

I tried to think from the point of view of a bush katydid. I swept diligently all the places which I suspected might appeal to a katydid. I tore my net on brambles, bayberry bushes, sensitive fern, bracken, arbor vitae, scrubby low-growing balsam, and vegetation in rocky places. In the process I swept up all the other species of bush katydids, together with a large enough variety of life forms to have stirred the envy of Noah. In the meantime, work at the laboratory continued during the day, and foraging during the early evening hours.

One moonless night, after an especially strenuous day, I celebrated by dining at the Franklin Café, as usual, and topped off the evening by attending an incredibly bad grade-B movie. I returned late, threaded my way through the woods toward the beacon of my cottage light, slipped through the door rapidly to thwart the hordes of mosquitoes (several abundant species), and stretched out for some bedtime reading.

I must have fallen into a doze, because it was late when I opened my eyes. I decided to call it a day. Usually I turned off the outside light once I was indoors. That night I had not done so. Well, I thought, I'll catch one breath of air before turning in for the night. When I stepped out into the persistent cloud of mosquitoes, I glanced casually at the light. On a shingle just at the edge of the penumbra clung a bush katydid. I took no chances. I turned slowly back into the cabin, picked up my net, reemerged as slowly, and with excessive caution clamped the net solidly and noisily against the shingle. The katydid did not budge. So much for my caution. I coaxed him gently into the bag and together we went inside.

The first thing I did upon removing him from the net was to look at his (for he was a male) tail. There was none! At long last I had a Northern Bush Katydid. He was very small for his genus,

Sweeping for bush katydids

only three-quarters of an inch long. This particular individual was even shorter than Forked-Tailed specimens I had previously caught. Thinking of the two, I was struck again by the difference in population density of the two species, the Fork-Tailed species very abundant, the Northern species rare.

That same night I established him in a small cage that I kept in the cabin for just such contingencies. I was even able to provide him with some lettuce, only slightly wilted, left over from the makings of that day's luncheon sandwich. Then, tired as I was, I turned out the lights and waited—and waited. Again I must have dozed, but sometime during the night I awakened to the sound I had so long sought. It was a soft train of ten long pulses lasting a fraction of a second and repeated every five or so seconds. It differed from the songs of all the other bush katydids. It was somewhat similar to the night call of the Broad-Winged species but was much softer.

The next morning I took him to the laboratory and broke the good news. We placed his cage in the photographic darkroom, hoping that he would be induced to sing; but his biological clock must have told him that this was not nighttime. Nevertheless, we left the door open a crack just in case he should sing. Around four o'clock he obliged us. Professor Pierce obtained six good recordings of the performance, probably the first ever recorded. We experienced the thrill of explorers accomplishing something that had never been accomplished before.

This was a small triumph compared to discovering a new river or ascending an unconquered mountain for the first time. It was not equal to discovering a new species; it was no momentous advance in scientific knowledge. In its own small way, it was an example of why men and women study nature. Of course, some people pursue science for fame or fortune or for some practical benefit to humanity. For others the pursuit is motivated by an irresistible urge to know, because lack of knowledge is a vacuum that must be filled. Human beings are the animals that must know.

Professor Pierce recording songs

This was the need that drove the old-time naturalists, a desire to understand nature for its own sake, not in order to manipulate it. As our culture becomes more sophisticated (or so we believe) and more utilitarian, we are apt to overlook much of the world of which we are a part.

These thoughts reminded me of the contrast between the definition of "entomology" in the eleventh edition of the *Encyclopaedia Britannica* and the official definition stated by the Entomological Society of America in 1966. Entomology as described by the encyclopedia is "the science that treats of insects." As stated by the society it is "the profession that controls insects."

So small a triumph as finding and listening to the Northern Bush Katydid brought to mind Edmund Burke's words in his *Reflections on the Revolution in France,* where he cautioned that those who make themselves most conspicuous are not necessarily the most important. We human beings are not necessarily the most important inhabitants of the planet. And even if we think we are, to study nature, to appreciate the world around us, if only to hear a katydid, is to have a more balanced perspective on our place in the cosmos.

Our katydid, even though we kept him indoors, lasted only till the first frost. We found him one morning lying stiffly on his side. Only then did I remove his wings to preserve them in balsam gum, as ancient insects had been preserved in amber. Then we were able to study his musical instrument, his file, and, however feeble his song, to understand it.

I never did discover that year where he had lived, where his consort abided, or whether there were others of his kind. In truth, I had never found him. He had come to me—or, more precisely, to my cottage lamp.

The Shield Bearers

Let every bough
Bear frequent vials, pregnant with the dregs
Of Moyle, or Mum, or Treacle's viscous juice;
They by the alluring odor drawn, in haste
Fly to the dulcet cates, and crowding sip
Their palatable bane.
John Phillips, "Cyder"

The stereotypical entomologist is an odd character, long of nose, short of vision, adorned with Ben Franklin spectacles, and given to dashing madly o'er the lea, with net and beard streaming in tandem in the breeze of his pursuit. This rather insensitive characterization has softened somewhat in modern times, as nonconformity has become conformity. The populace still retains, however, one illusion regarding "bug hunters": it is that the chase is all storm and fury and that success goes to the fleetest. I have tried to dispel that illusion in this account of our studies of singing orthopterans, by describing stratagems and tactics, the need for stealth, guile, and patience, and the value of intelligence, in the sense of military intelligence. I have alluded to the importance of understanding idiosyncrasies of the quarry's behavior, selecting appropriate habitats for exploration, choosing the opportune hour and season. Above all, one must court the muse of serendipity.

As the search for the Northern Bush Katydid revealed, adherence to every one of these principles may still be perversely unrewarded. One may be defeated by the paradox that nature is at one and the same time structured and chaotic, orderly and random. One needs, therefore, a modicum of blind luck, the chance of being in the right place at the right time. One does not always know where the right place is. The fact that a species has not been reported from a particular area is in itself no indication that the insect does not occur there. Collectors are inclined to avoid such sites and overexplore those with reputations for abundance. Thus, the unexplored spots acquire undeserved reputations for scarcity. They may be just the places to search.

Another gambit that may seem counterintuitive is the knack of bringing to bear on the problem seemingly irrelevant knowledge or knowledge which at best seems only peripheral to the situation. The capture of the Northern Bush Katydid is a case in point. If one is hunting moths, there is a precedent for employing light as a lure. It had never occurred to me that some katydids might also be attracted. I had not, therefore, set up any light traps. Now, in my search for shield bearers, the lesson of irrelevance put me on the track of a new stratagem. The initial event in a chain of circumstances which led finally to success was a badminton game.

The Cannons had laid out a badminton court on the lawn fronting the house. There they, their friends, and their visitors played at the tag end of the day or early evening, until the afterglow no longer provided enough light for following the bird. On occasion, when a fourth was needed to play a doubles match, I was invited to participate. Beyond that I was always welcome to come as a spectator and occupy one of the reclining chairs along the sidelines. This social gathering alleviated the solitude of routinely retiring to my cabin or going bowling in town yet again.

As a player I occasionally found myself teamed with a prestigious elder, or playing against one. In the latter circumstances, I once had the dubious distinction of driving the bird into Professor Cannon's face, an incident I could never have foreseen during my student days as a lowly attendee at his lectures. His lectures had been more thoughtful than inspiring, but his books more than made up for any disappointment I may have suffered. Three of his books (*Bodily Changes in Pain, Hunger, Fear, and Rage; The Wisdom of the Body;* and *The Way of an Investigator*) had been assigned tutorial read-

ing in my undergraduate days and, although I did not realize it at
the time, they helped direct my own career into experimental
physiology.

In the less hazardous role of spectator, I enjoyed the privilege
of listening to conversations that ran the gamut from sociological
problems of the period and the politics of Roosevelt, to the pros
and cons of the Spanish Civil War, to wartime medicine. Professor
Cannon, who for humane reasons had helped organize medical
assistance to the Spanish Republic, was being roundly criticized
for his beliefs by isolationists and Fascist sympathizers in the
United States. Among his friends and close associates, his human-
itarian motives were understood and his perseverance in the face
of hostility admired. To me, these erudite and congenial concerns
were a welcome respite from daily preoccupations with orthopterans. Despite my best intentions, however, irritation at the elu-
siveness of the shield bearer still intruded, at least subconsciously.

One evening, we had just finished a doubles game in which
Professor Pierce and I were handily beaten by Professor Cannon
and one of his daughters. We agreed that it had become difficult
to see the bird in the fading light. Play was terminated, but nobody seemed inclined to retreat indoors. The conversation
dropped to a quiet murmur of reminiscences. This was the kind of
night evoked in *The Merchant of Venice*:

> Here we will sit and let the sounds of music
> Creep in our ears: soft stillness and the night
> Become the touches of sweet harmony.

Relaxing into a state of inattention, I gradually became aware
of more than stillness. The silence was delicately accentuated by
faint night noises muffled by distance. There were not many. Calls
of bush katydids drifted up from the direction of the fields and
hedges. Even in their softness, I realized that there was nothing
novel there, so I welcomed them only as one element of a romantic setting to the night. In the immediate surroundings all was

American and Short-Legged Shield Bearers

quiet. No cricket stirred in this habitat. Behind the house, where the studio was located and the woods were mostly pine, no sound broke the stillness except for a brief incoherent peep from a bird stirring in his sleep.

Without turning my head, I could direct my auditory attention toward the woods on the far side of the badminton court. There an open stand of oaks and maples was silhouetted against a blue-black sky. At one point I thought I heard a soft sound from that direction. It could have been a bough creaking, but there was not enough wind to stir a single leaf. When the sound was repeated, I wanted to investigate but could not face the embarrassment of abandoning the group in order to poke among the trees in the dark. It sometimes takes a greater measure of sangfroid to be a good entomologist than I was able to muster at that moment. I made some concession to social graces by yanking my attention back to the conversation around me. When the group did at last begin to disperse, I mentioned to Mrs. Cannon that I might, if she acquiesced, return later with a flashlight. She understood.

About ten o'clock I did return to the woods. There were indeed distinct sounds emanating from somewhere among the dry leaves that carpeted the ground. I tried to sneak up on the source of the sound but in the process made so much rustling that the sound maker stopped. I knew then that some animal, and not a tree, was creaking and that it was most probably an insect. At one point I zoned in on a soft lisping buzzing, but it ceased abruptly at my approach. In the ensuing silence I heard a faint rustling in the leaves as of something hopping, something more light-footed than a wood frog but less ethereal than an elf. A quick sweep of the light beam revealed nothing. An hour of searching and tracking similar sounds passed unrewarded. Shafts of light from a full moon were of little help. At midnight I abandoned the quest.

Returning to the cabin, I glanced perfunctorily at the light. Nights with a full moon are generally poor times for trapping by light, but there were some insects abroad. A few nondescript moths, some craneflies, a single longhorn beetle, and a pair of ichneumon wasps circled about or rested on the shingles, possibly to regain their senses after having banged their heads against the globe. As I watched, a few more moths caromed out of the darkness. The light that attracted this assemblage had ten days earlier lured the Northern Bush Katydid, but tonight there were none of his brethren.

At that point I made the irrelevant connection: moths were attracted to light; light had attracted the katydid; moths were also attracted to sweet baits. Perhaps sweet baits placed on the ground would attract shield bearers, if indeed they were the sources of the footfalls in the leaf litter.

The next morning I followed my hunch and was rewarded when I rummaged through some old entomological literature. There I found a mention of shield bearers' being trapped in cans baited with molasses and sunk in the ground.

At lunchtime I visited Surowiec's Market. Mr. Surowiec did not stock the Moyle, Mum, or Treacle of eighteenth-century recipes, but he did have the equivalent: beer and blackstrap molasses. I contented myself with beer and molasses. From these I mixed a savory concoction, to which I added some cornmeal mush. The latter would provide a soggy substrate so that any unwary tippler who missed his footing would not drown in his drink, a circumstance that had inspired several seventeenth- and eighteenth-century English poets to write about insects and spiritous liquors. Richard Lovelace had mused about a fly attracted to a glass of burned claret. John Walcott had chronicled the fate of a fly rescued from a bowl of punch.

That very afternoon I sank several of my sweet alcoholic traps in the ground among leaves of the hardwood forest. Each night thereafter I visited the cans and removed any leaves that had blown in, together with an assortment of tipsy fauna.

As the nights passed, the brew, with the assistance of ambient yeast spores, became more alcoholic, more redolent, and, judging from the increasingly large number of patrons removed at each visit, more alluring. On the fifth night, a moonless humid night, I found that which I had so long sought. In two of the traps there were not only shield bearers of both sexes but shield bearers of both species. At my approach they made no attempt to escape from the cans. Even if they had been sober, it is doubtful that they could have climbed the sheer sides. Shield bearers are not very active jumpers under any circumstances.

Nor had they been fastidious tipplers. In fact, they were ludicrously sticky from having walked around on the bait and upon one another.

Whether to attempt to clean them or leave them to make their own toilettes was resolved when I transferred them to a small cage. There each began to clean his or her antennae by using the front legs to bring the antennae one at a time into the mouth, where each was threaded through from one end to the other. The sight reminded me of a person eating corn on the cob. Each leg was then cleaned in the same manner. This foot-in-mouth performance convinced me that the insects would do a better job of grooming themselves than if I attempted the dainty task with my clumsy fingers.

Most animals are very conscientious about keeping themselves clean, and are more attentive to cleanliness than some human beings. Insects are no exception. For them survival depends upon cleanliness, because the majority of their sense organs are hairs, each adaptively constructed for a special function. Taste is served by hairs on the mouthparts and legs. The antennae are equipped with hairs to detect the sweet aromas of flowers, the alluring aphrodisiacs of the opposite sex, the scented homeward path to the nest, the meaty odor of prey, or the warning effluvia of the unpalatable. An equally exquisite sense, the sense of touch, also depends on hairs of various types scattered everywhere over the body. To avoid losing contact with the world around it, an insect must be well groomed.

So important is grooming that it takes priority over most other behaviors. It is neither learned nor perfected with practice. It is genetically built into the insect for immediate use from the beginning of adult life. The neural machinery is so designed that no brains are required. A headless fly can still groom its body.

My shield bearers eventually cleaned themselves to their own satisfaction. Before long they were contentedly (I imagine) feasting on lettuce, plus an occasional tidbit of apple. The older the apple the more appetizing it appeared to be.

I had separated the catch by species into male/female pairs. I feared to keep them as colonies, because shield bearers have been reported as being disgracefully cannibalistic when crowded or confined under suboptimal conditions. The incidence is reduced if cages are kept moist. In some circumstances cannibalism is driven by a need more for water than for food. As isolated pairs, my captives performed well.

The two species are much alike in behavior but easily distinguishable in appearance. The Short-Legged Shield Bearer possesses hind legs that are one and one-half the length of the body; the hind legs of the Long-Legged Shield Bearer are twice as long as the body. Aside from this difference, the two are similar. They are about the size of field crickets. Both are chunky, angular, brown, cricketlike insects that look like toy armored vehicles. The chunky appearance is accentuated by the broad faces, small, widely separated eyes, and very short wings. The back of the thorax extends posteriorly as an enlarged rigid projection which covers the first segment or two of the abdomen. This "shield" broadens at the back and arches over the bases of the wings. It almost conceals the file and the scraper. One wonders how the wings can produce their calling song in such a restricted space.

That song is not spectacular, in any case. The short-legged species, the less common of the two, calls with a series of repeated trains of z-e-e-e, each lasting from three to three and one-half seconds. This song resembles that of the Common Meadow

Grasshopper. There is no confusing the two, because the zips of the grasshopper song are absent. Furthermore, the grasshopper inhabits tall grasses in open areas, whereas the shield bearer is a habitué of upland open wooded areas. Less frequently the shield bearer elects to live in brushy pastures. The short-legged species inhabits much the same terrain; he calls with a song consisting of shorter trains of pulses or buzzes, each lasting less than half a second.

Having solved the riddle of the sounds from the far side of the badminton court, I was able to put the matter out of my mind. All noises had been accounted for, for the moment. Our collection of local insect songs was nearly complete. But there was another aspect of the sense of fulfillment. The songs were part of the visage and soul and totality of nature. With more complete understanding, there comes an opportunity for deeper appreciation. On the other hand, perhaps one should not know nature too well; there is allure in mystery. We had seen and heard facets of nature that enhanced our enjoyment. Enough mystery remained—and would remain, so long as the itch of curiosity kept us restless.

This was to be the last subdued interlude. Already the waning summer was bringing a new change, as another group of insects approached their musical apogee. We were about to experience an extraordinary coda to the summer symphony.

Déjà Entendu

I see thee quaintly
Beneath the leaf; thy shell-shaped winglets faintly—
(As thin as spangle
Of cobwebbed rain)—held up at airy angle;
I hear thy tinkle
With faery notes the silvery stillness sprinkle—
Madison Carwein, "The Leaf Crickets"

Mid-September was upon us before we fully realized it. Labor Day had come and gone, as had most of the human summer residents. Mercantile activity in town had settled into the more measured pace that would sustain it throughout the long winter. The town had drawn back into itself almost as though summertime had been an interlude, an interlude provided for the sole purpose of preparing for the nonproductive months ahead.

Our stay at the laboratory was now measured in days. We were caught between an unsettling sense of urgency and an elated sense of expectancy—urgency because we had only two weeks in which to complete our work, expectancy because the last group of insects to be studied, the tree crickets, had just reached the peak of their activity. All about us, there was marked change.

Mornings were cooler and crisper than before. The ever-lengthening shapes of afternoon shadows seemed drawn more irresistibly into the night. Fields were rough and tweedy, as though an old brown woolen jacket had been thrown over them to ward off the chill. Along roadsides, in abandoned pastures, on open hillsides, hues had turned sombre. The flowers of fall lacked the brilliance and freshness of their spring predecessors. Their colors were the colors of the aged. One saw lavender and old lace: the violet-purple of New England Asters alongside the faded white of

Umbellate Asters. Queen Anne's Lace had lost its virginal white-
ness and gone to seed. In seeming paradox, the once brilliant
goldenrod appeared tarnished. Even the textures of plants were
those of the aged. Vegetation which before had been soft, smooth,
and supple had become rough, coarse, and brittle.

How incongruous that this hoary backdrop should be the
setting for the performance of the most jubilant of insect choruses!
No more incongruous, perhaps, than is the contrast between the
delicacy of the performers and the robustness of their music. Tree
crickets have been described by Morse as "veritable dryads of
fairy-like daintiness and evasiveness, often heard but seldom seen."
Thoreau described the song of the Snowy Tree Cricket as a "slumbrous breathing," an "intenser dream." Hawthorne called it "an audible stillness" and added that "if moonlight could be heard, it
would sound like that." It is small wonder that I was anxious to see
these crickets and listen to their songs at close range.

The first male that I caught had been singing in late afternoon from a perch in a stand of goldenrod. Locating him proved
to be more difficult than I had anticipated. Even though the whole
field was populated with myriad singing tree crickets, I had no
trouble singling out a particular clump containing one cricket. Exactly where in the clump he was perched was quite a different
matter. These crickets seemed to be ventriloquists. When I stood
to one side of the clump, the cricket seemed to be on one particular plant. When I moved to another side, he seemed to be in a
different plant. As I moved around trying to get a fix, I had an
overwhelming conviction that he, too, was cunningly moving
around away from me. Eventually I spotted him. As an experiment,
I changed my position several times, always keeping him in sight.
Throughout this game of hide-and-seek he continued singing
lustily, without stirring from his perch; nevertheless, as I moved,
his song seemed to come from a different locus.

The song was clearly ventriloquial, but the cricket was not a
ventriloquist; that is, he himself was not producing sound in such

a way as to make it appear as though it were coming from another source. The trick was played by the surroundings, in which a song of a certain frequency and harmonics was reflected, refracted, and absorbed by different stems, leaves, and clumps of vegetation. He was singing in a concert hall of complex acoustics where there were different echoes, whispering spots, and dead spots, depending on where one sat in the hall. Different acoustic environments play different tricks with different songs, but the effect on the listener is always eerie and frustrating.

From whatever point the sound was heard, it was glorious. The musician himself was so exquisite that I watched for a long time before disturbing him. One has to hear the song for oneself to appreciate its musicality. The singer was, as the poets have described him, a "veritable dryad of fairy-like daintiness" with "shell-shaped winglets as thin as spangle." In more prosaic terms, he was a fragile, pale yellowish-green creature with flat, broad, diaphanous wing covers. As the cricket sang, the wing covers were kept raised at a forty-five-degree angle and appeared blurred with vibration. The song, which was a musical trill that seemed to continue interminably, from late afternoon throughout the night, consisted, technically speaking, of single pulses strummed at the rate of fifty per second.

After listening entranced for several minutes I scooped him up, decapitating the goldenrod in the process and netting also a female that had been as mesmerized by the song as I. She was easily distinguishable from the male because her slender wings rested lengthwise, closely appressed along the body.

In the laboratory the next day, the crickets were subjected to a close but gentle examination. Tree crickets are most easily identified by black markings on the basal segment of the antennae. These had two such marks, which were so broad that the segment was almost entirely black. Furthermore, the back was black. The markings clearly identified the specimens as Black-Horned Tree Crickets.

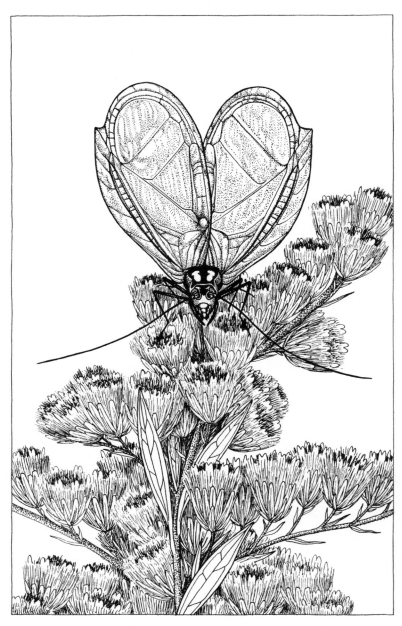

The Black-Horned Tree Cricket

That afternoon I returned to a different part of the field, where the weedy plants were less coarse and did not exceed three feet in height. Here also there must have been hundreds of tree crickets singing. In chorus the songs resembled those of the Black-Horned species; nonetheless, I decided to sweep the weeds. The first sweep yielded two males and a female. They seemed lighter in color than those caught the day before. Only later was I able to determine that they were specimens of the Four-Spotted Tree Cricket. The basal segment of the antenna had two spots, as did the next segment. The back was not black. The song resembled that of the darker species. Upon analysis it was found to consist of double rather than single pulses and to average slightly fewer per second (average forty-one) than that of its black-horned relative.

According to the books, there were three other species that we might have expected to find: the Narrow-Winged, the Pine, and the Snowy Tree Crickets. We never did find any members of the Narrow-Winged species. On my trip to Cape Cod I had captured one about twenty-five feet up in a scrub oak (a feat of pure luck), but in the laboratory in Franklin he had refused to sing.

Eventually I did locate some males of the Pine Tree Cricket, easily identifiable because they are partial to pine trees. Of the two spots on the basal antennal segment, one is very indistinct. The song, like that of the others, is a continuous trill, rather soft and silvery. The trill is single-pulsed.

The Snowy Tree Cricket is easily distinguished from the other species that we studied. The male has much broader wings, which make him appear less svelte. On the other hand, his very pale coloring, almost white, gives him a more ethereal appearance. Since he has these obvious characteristics, it is not necessary to examine the base of the antennae, where a single spot confirms his identity.

Insofar as song is concerned, the Snowy Tree Cricket is the most evocative musician of the group. Literary descriptions are not too fanciful when they refer to "a slumbrous breathing" and "the sound of moonlight, if that could be heard." His song is a rhythmic musical chirping of pulsed trills, two or three to the second.

One never tires of watching these exquisite insects and is eternally impressed by their delicate form and soft coloration. How such refined creatures can produce the volume of sound they do is further cause for wonderment.

The answer to the puzzle of loudness is not to be found in the amount of energy expended, the level of motivation, or the robustness of the player. The secret lies in the structure and constitution of the wings. Scrapers bumping over files do not produce loud sounds. A person has only to run his thumbnail across the teeth of a comb to notice the low volume of sound. It is only when the vibrations are transmitted to some kind of a sounding board that the sound is enhanced and transformed. Again one can demonstrate this for oneself by scraping a comb and at the same time holding it against some substrate that will resonate. The wings of the insect are its resonating boards. Their uneven structure allows for many harmonics. Some wings even have miniature drumheads, which cause a more pronounced radiation of sound of a frequency that is resonant with the disc's natural frequency. In some insects there are additional refinements. The wings of meadow grasshoppers and katydids, when positioned for file and scraper to engage, form a trumpet in which the column of air resonates, thus further enhancing the sound.

The insect has minimal control over these matters. He can, within limits, play different variations. He can vary speed and sequences to produce calling songs, courting songs, and songs of aggression. He can modulate sound slightly, but he has no control over pitch, tone, timbre, color, and mode of expression. There are no virtuosi among performers. The instrument is everything. Each player is born with either a Stradivarius or a cheap box.

It is perhaps unfeeling to say so, but the fact remains that the players are uninspired automatons fated by genetics and their hormonal tides to be what they are when they are. But knowing this does not diminish the romance that the sounds evoke. Even as a good music box can give pleasure and affect moods and memories, so can files and scrapers if they happen to be constructed in a way that, serving their natural purpose, coincidentally pleases the human ear and, further, affects the human mood when played in felicitous circumstances.

Loudness is, of course, more than the accomplishment of a single player. There is also the size of the orchestra to be considered. Some insects are soloists performing alone on the ground, in crevices, under leaves, in clumps of grass, in tangles of weeds, or in aeries in bushes or trees. They may be few in number, rare, restricted to small areas, or widely spaced across broad reaches of field or forest. Others, though nonsocial, may be packed together in such densities as to make a field resonate with their trills and tinkles. Many species of ground crickets make up such assemblages in fields, pastures, and marshes. Meadow grasshoppers form comparable assemblages. Tree crickets, especially the Black-Horned and Four-Spotted species, also may congregate in such numbers as to make a swale or hillside ring with their trills. In all these cases the activity of one individual encourages others to sing. There are few orthopterans, however, that sing in such closely synchronized choruses.

Snowy Tree Crickets are choristers par excellence. One male starts his song, then another, then the next and the next. Each one starts a fraction of a second after his predecessor, but the delay is so short as to be undetectable by the human ear. The result, for us, is a huge synchronized chorus. When large numbers of crickets make up these choruses, the hedgerows, orchards, and shade trees resound.

These songs begin a few at a time late in August, swell in number and volume in September, and continue into October. They overwhelm the other songs partly because of their volume, partly because the tree crickets sing from places of vantage, and possibly because we who have listened to other singers from early summer give more of our attention to what is novel.

Since these choruses come so late in the season, it seems almost as though song might stay the passage of time. At this point, one succumbs to the illusion that this shrilling chorus has been heard before, in another time and another place, a time of commencement and a place of chilling water. And then the sound and the picture fall into place. The shrilling of the tree crickets is a sonic déjà vu, a déjà entendu, of the chorusing spring peepers in the swamps and bogs. The two choruses are remarkably similar, not only in fancy but in sound and pitch and rhythm. The resemblance brings the listener to a rude realization of the passage of time. Only yesterday it was spring. Today it is fall. The year is ending on the same note on which it began.

Reprise

I know the night is now at hand.
The mists lie low on hill and bay,
The autumn sheaves are dewless dry;
But I have had the day.
Silas Weir Mitchell, "Vesperal"

Toward the end of September, with the fields and woods still echoing with song, we released into their home habitats those insects still in our custody, closed the laboratory, and transported the records back to Cambridge. There, at Harvard's Cruft Laboratory, Professor Pierce would continue his analyses and Paul Donaldson would prepare photographs for eventual publication. I would complete studies of pinned specimens that had been collected during the summer and would verify my identifications by comparison with specimens in the splendid collection at Harvard's Museum of Comparative Zoology. We had made no plans for further recording of songs. Events were, however, to dictate otherwise.

Scientific research is somewhat like a banquet. One accumulates data, food for thought, after which comes digestion and assimilation. Sometimes the sequence is interrupted by presentation of an especially appetizing dessert. Our dessert, donated by a colleague at the Boston Museum of Natural History, was a live male True Northern Katydid. This was a cherished gift indeed, because these raucous, argumentative insects reside in small colonies high up in trees, where they are safe from marauding entomologists. Often heard but seldom seen, they are elusive, secretive, and silent during the day. They serenade only at night, continuously from dusk till dawn. The male captured by our colleague had fallen from his tree, either blown from an exposed perch by a caprice of the wind or beaten earthward by a brief rain squall. His misfortune was our good luck, especially because the true katydid does not range as far north as Franklin.

True katydids are one of the few members of their family that thrive in the neighborhood of human beings. When early colonists cleared the land, laid out their villages and towns, and planted shade trees along the streets, the katydids became members of the community. Few New England villages were without their katydids or their crickets on the hearth. Both insects were reminders of the greater nature of which mankind was but a part.

With the coming of fall, katydids take possession of the maples, lindens, and elms which in the sultry dog days of summer were the undisputed domain of the cicadas. It is fitting that as the season of warmth and growth approaches its end, the singing should be reserved for night, whereas earlier in the year the cicadas chorused during the day. It seems appropriate also that the performers of summer should revel in the heat and sing with frenzy, whereas the katydids sang with sonority in the coolness of fall.

In earlier times, colonies of katydids were more numerous in the shade trees of highways and byways in villages, towns, and suburbs. All that changed when the spraying of insecticides to control gypsy moths and Japanese beetles left the crowns of trees barren and dripping with chemical rain. Sprays are no respecters of species. Many colonies of katydids succumbed and left no progeny. A distinctive part of the atmosphere of New England towns and villages became in some places only a memory.

When one hears katydids, one is transported back in memory and, by extrapolation, back into history, to times when the first trees were set out along the dirt road that was Main Street and that later became its stately paved equivalent, as the towns aged and the cities reached ever farther into the country. By establishing towns, humans created a habitat comfortable for them-

The Northern True Katydid

selves and congenial enough to lure robins, swallows, orioles, and katydids from the countryside. No human community has ever been complete without its neighbors from the fields and copses. Most of these neighbors have been celebrated in folklore and verse.

The katydid's first librettist was probably Philip Freneau, who lived in the eighteenth century and who was called the "first national poet of America."

> In her suit of green arrayed,
> Hear her singing in the shade—
> Caty-did, Caty-did, Caty-did!

It surely must have been poetic license that prompted Freneau to interpret the insect's song as "Caty-did" (the original spelling). It is difficult for most of us to associate the sound of the song with those words. Furthermore, the tone is neither romantic nor musical. Nevertheless, it is a homely sound that one associates with quiet, peaceful, shady neighborhoods. When uttered by a single individual, it is a loud, hoarse, rasping, grating phrase repeated interminably throughout the night. But it belongs.

Katydids seem to rejoice in answering one another. They respond equally enthusiastically to imitative artificial sounds. In the laboratory, when we made a two-syllable sound, the katydid responded in kind. If we made a noise of three pulses, the response was three; if four, the response was four. None of the males seemed inclined or able to count beyond four.

Although individual songs are unattractive to the human ear, a chorus of katydids does possess a certain charm. Sometimes each male will sing in alternation with the nearest neighbor and he in alternation with another, so that with large numbers of singers on calm humid nights there is grand choral synchrony. If some males sing three-pulse songs while others sing two-pulse songs, the chorus may be disrupted and a curious medley may result.

Freneau, in ascribing the song to the female katydid, erred;

only the males sing the refrain. The poet's verse may also have been incorrect in another sense, because it is unlikely that Katy herself would proclaim that she did whatever it was that had been done. The females do produce sounds under appropriate circumstances—they are one of the few female orthopterans that do. They respond to calling males with a squawk that is anything but dulcet; nevertheless, the male responds in turn.

Shortly after the colonies in the shade trees were stilled by autumnal chills, our pet katydid died. The tree crickets and their predecessors were already gone. The music of the last ground crickets to sing, the stuttering Carolina Ground Crickets, faded into silence as the nights grew colder. The crickets predeceased our katydid by a few days. Their departure reminded me of Haydn's Farewell Symphony, in which, as the last movement progresses, the musicians leave the stage one by one, each extinguishing the light on his music stand, until only two violinists remain.

By the end of Indian Summer, the silence that was to be winter had muted the land. Few sounds were heard from the animate world. Bluejays were vocal as always; they changed their tune from a nasal "Thief! Thief! Thief!" to the more melodious yodel of fall. Occasionally, on a late afternoon, the honking of the first chevrons of southward-bound Canada geese would drift earthward from a lowering sky. The only other sounds were the aeolian whispers, sighs, moans, and wails of the wind as it played on the wizened stalks of sere fields, the bare crowns of woodland trees, and the cornices and loose shutters of houses. Gradually I adjusted to the passing of summer and to my return from the disordered beauty of rural habitats to the manicured landscape of suburbia. I was not prepared for a reprise.

Three days before Thanksgiving, as I was hurrying through Harvard Yard, I heard a single cricket chirping. The chirp sounded exactly like that which in early June had ushered in a summer of song. This time, however, the caller was a Fall Field Cricket that had found temporary shelter near the grating of a heating vent outside Thayer Hall. Whereas the chirping of the Spring Field Cricket had resonated with promise, had evoked memories of summers past and anticipation of summers to come, this song, with identical scoring and execution, evoked different emotions. Even as I let my imagination range, I realized anew how much the listener brings to the music, how music evokes moods complementary to its setting, and how moods close the circle by shading the music. I felt a sense of melancholy, listening to the cricket. He was calling, and there was no mate to listen. He was calling into the void of imminent winter. Yet in that melancholy, I experienced—if not anticipation and assurance—at least hope for another spring.

Two days later it snowed.

Appendixes
References

Species Studied at the Franklin Laboratory

Names in parentheses indicate names in use at time of study

Field Crickets
Northern Spring Field Cricket	*Gryllus veletis*
Northern Fall Field Cricket	*Gryllus pennsylvanicus*

Ground Crickets
Allard's Ground Cricket	*(Nemobius fasciatus) Allonemobius allardi*
Striped Ground Cricket	*(Nemobius socius) Allonemobius fasciatus*
Sand (Gray) Ground Cricket	*(Nemobius griseus) Allonemobius griseus*
Tinkling Ground Cricket	*(Nemobius tinnulus) Allonemobius tinnulus*
Carolina Ground Cricket	*(Nemobius carolinus) Eunemobius carolinus*
Sphagnum Ground Cricket	*(Nemobius palustris) Neonemobius palustris*

Meadow Grasshoppers
Common Meadow Grasshopper	*Orchelimum vulgare*
Gladiator Meadow Grasshopper	*Orchelimum gladiator*
Dusky-Faced Meadow Grasshopper	*Orchelimum concinnum*
Slender Meadow Grasshopper	*Conocephalus fasciatus*
Short-Winged Meadow Grasshopper	*Conocephalus brevipennis*
Salt-Marsh Meadow Grasshopper	*Conocephalus spartinae*

Coneheaded Grasshoppers
Sword-Bearer Conehead	*Neoconocephalus ensiger*
Robust Conehead	*Neoconocephalus robustus*

Bush Katydids
Oblong-Winged Katydid	*Amblycorypha oblongifolia*
Round-Winged Katydid	*Amblycorypha rotundifolia*
Curve-Tailed Bush Katydid	*Scudderia curvicauda*
Fork-Tailed Bush Katydid	*Scudderia furcata*
Broad-Winged Bush Katydid	*Scudderia pistillata*
Texan Bush Katydid	*Scudderia texensis*
Northern Bush Katydid	*Scudderia septentrionalis*

Shield Bearers
    Long-Legged Shield Bearer   *Atlanticus americanus*
    Short-Legged Shield Bearer   *Atlanticus testaceus*

True Katydids
    Northern True Katydid   *Pterophylla camellifolia*

Tree Crickets
    Black-Horned Tree Cricket   *Oecanthus nigricornis*
    Four-Spotted Tree Cricket   *Oecanthus quadripunctatus*
    Narrow-Winged Tree Cricket   *Oecanthus angustipennis*
    Pine Tree Cricket   *Oecanthus pini*
    Snowy Tree Cricket   *Oecanthus niveus*

Slant-Faced Locusts
    Marsh Meadow Locust   *Chorthippus curtipennis*
    Sprinkled Locust   *Choealtis conspersa*
    Bunch Grass Locust   *Pseudopomala brachyptera*
    Pasture Locust   *Orphulella speciosa*

Band-Winged Locusts
    Spring Sulphur-Winged Locust *Arphia sulphurea*
    Autumn Sulphur-Winged Locust *Arphia xanthoptera*
    Clouded Locust   *Encoptolophus sordidus*
    Black-Winged Locust   *Dissosteira carolina*
    Ledge Locust   *Spharagemon saxatile*
    Cracker Locust   *(Circotettix verruculatus) Trimerotropis*
                                     verruculata

Seasonal Distribution of Adult Singing Orthoptera in Franklin, New Hampshire

Broken line indicates that killing frost determines end of season

May	June	July	August	September	October	November

├────── *Gryllus veletis* ──────┤

├──────────────────── *Arphia sulphurea* ──────────────────── - - - -

├────────── *Chorthippus curtipennis* + *Chloealtis conspersa* ────────── - - - -

├────────── *Allonemobius fasciatus* + *allardi* ──────────┤

├──────────────── *Dissosteira carolina* ──────────────── - - - -

├──────────────── *Arphia xanthoptera* ──────────────── - - - -

├──────────── *Gryllus pennsylvanicus* ──────────────── - - - -

├──── *Pseudopomala brachyptera* ────┤

├────────── *Orchelimum* species ────────── - - - -

├──── *Conocephalus fasciatus* + *brevipennis* ── - - - - -

├──── *Conocephalus spartinae* ────┤

├────────── *Spharagemon saxatile* ──────────┤

├────────── *Trimerotropis verraculata* ────────── - - - -

├──── *Scudderia* species ────┤

├────────── *Atlanticus* species ──────────┤

├────────── *Encoptolophus sordidus* ────────── - - - -

├ *Neoconocephalus* species ┤

├────────── *Oecanthus* species ────────── - - - -

├────────── *Allonemobius tinnulus* ────────┤

├ *Allonemobius griseus* ┤

├────────── *Neonemobius palustris* ────────── - - - -

├────────── *Eunemobius carolinus* ────────── - - - -

├ *Amblycorypha* species ┤

├────────── *Pterophylla camellifolia* ────────── - - - -

Key for Identifying Species Studied

The following keys were designed to assist the nonspecialist in separating and identifying the singing Orthoptera described in the Franklin study. Only those characters which can be observed without the use of a microscope are used. For more technical and comprehensive keys covering a wider geographic range, the reader should consult the works listed under "References," especially the paper by Alexander, Pace, and Otte.

To use this key, begin at number 1 and select 1 or 1'. If 1 is the proper description, proceed to 31. If 1' is the proper description, proceed to 2. Continue selecting pairs until a species name is given.

1	Antennae less than one-half body length, threadlike, and somewhat stubby (locusts)	31
1'	Antennae longer than body, filamentous (crickets, meadow grasshoppers, katydids)	2
2	Wing covers vertical (meadow grasshoppers, katydids)	14
2'	Wing covers horizontal (field crickets, ground crickets, tree crickets)	3
3	Body color yellowish-green to green	11
3'	Body color brownish-black to black	4
4	Small (5.5–11 mm)	6
4'	Large (13–21.5 mm)	5
5	Adults from May to July	
	Northern Spring Field Cricket (*Acheta veletis*)	
5'	Adults from August to frost	
	Northern Fall Field Cricket (*Acheta pennsylvanicus*)	
6	Reddish-brown to black	7
6'	Reddish-brown with gray pile, in sandy areas	
	Sand or Gray Ground Cricket (*Allonemobius griseus*)	
7	Larger than 6 mm	8
7'	Very small (5–6.2 mm), jet black, in bogs	
	Sphagnum Ground Cricket (*Neonemobius palustris*)	

20   Heads narrow 22

20' Heads round 21

21   Forewings 30–40 mm long . . . .
. . . . Oblong-Winged Katydid (*Amblycorypha oblongifolia*)

21' Forewings 23–27 mm long . . . .
. . . . Round-Winged Katydid (*Amblycorypha rotundifolia*)

22   Heads pinched between antennae 27

22' Heads slanting 23

23   Not cone-shaped 25

23' Acutely cone-shaped 24

24   Large body (51–56.5 mm) salt-marsh species . . . .
. . . . Robust Conehead (*Neoconocephalus robustus*)

24' Smaller body (45–50 mm) . . . .
. . . . Sword-Bearer Conehead (*Neoconocephalus ensiger*)

25   Large (20–30 mm), robust 26

25' Smaller than 17 mm, slender . . . .
. . . . Slender Meadow Grasshopper (*Conocephalus fasciatus*)

26   Green . . . .
. . . Common and Gladiator Meadow Grasshoppers (*Orchelimum*
vulgare and *O. gladiator*)

26' Dusky . . . .
. . . . Dusky-Faced Meadow Grasshopper (*Orchelimum concinnum*)

27   Last dorsal segment of male with forked notch 28

27' Last dorsal segment not prolonged with notch or fork . . . .
. . . . Northern Bush Katydid (*Scudderia septentrionalis*)

28   Fork U-shaped or V-shaped, no central tooth 29

28' Fork broad, with central tooth . . . .
. . . . Texan Bush Katydid (*Scudderia texensis*)

29   Notch V-shaped 30

29' Notch U-shaped, like small hoof . . . .
. . . . Fork-Tailed Bush Katydid (*Scudderia furcata*)

30   Fork, viewed from above, narrow and tapering . . . .
. . . . Broad-Winged Bush Katydid (*Scudderia pistillata*)

30' Viewed from above, rounded . . . .
. . . . Curve-Tailed Bush Katydid (*Scudderia curvicauda*)

31   Face rounded, not slanting backward . . . .
. . . . Band-Winged Locusts 35

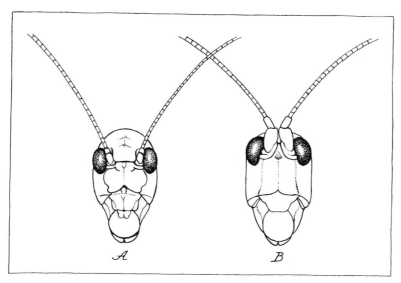

Faces of round-headed katydid (A) and bush katydid (B)

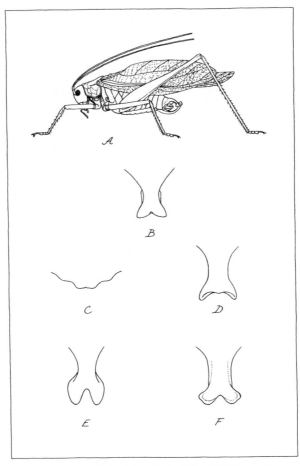

*"Tail plates" of male bush katydids. (A and B) S. pistillata;
(C) S. septentrionalis; (D) S. texensis; (E) S. furcata; (F) S.
curvicauda.*

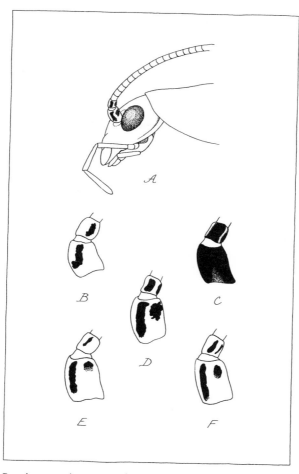

Basal antennal segments of tree crickets. (A, C, and D) O.
nigricornis; (B) O. niveus; (E) O. pini; (F) O. quadripunc-
tatus.

37′ Adults in August to frost. . . . .
     . . . . Autumn Yellow-Winged Locust (*Arphia xanthoptera*)
38   Hind wings without black spur to wing base 39
38′ Hind wings with black line to base of wings, hind tibial fuscous at
     base, followed by white ring, red distally, body with brown-
     black spots on ash-gray ground. . . . .
     . . . . Ledge Locust (*Spharagemon saxatile*)
39   Hind tibiae yellowish-white, blackish at base and apex, wing disc
     very pale yellow. . . . . Cracker Locust (*Trimerotropis verruculata*)
39′ Hind tibiae brown or black, body dull brown or gray, wings faintly
     clouded, dusky at tip . . . .
     . . . . Clouded Locust (*Encoptolophus sordidus*)

Phonetic Key to Songs of Crickets, Locusts, and Meadow Grasshoppers

Song keys based on phonetics are notoriously subjective and songs really must be heard to become recognizable. The key is, therefore, only an approximation of the sounds as interpreted by different observers. It will help separate the major groups (crickets, locusts, meadow grasshoppers, and katydids). For species identification, one must combine song, habitat, and seasonal characteristics and ultimately capture and study the insect to determine morphological features. Once heard, the songs will be recognizable and remembered.

1	Clear, musical chirps or trills . . . . *(crickets)* 2	
1'	Nonmusical buzzes, clicks, whispers, snaps.	
	. . . . *(locusts, meadow grasshoppers, katydids)* 11	
2	Chirps or tinkles. 3	
2'	Trills. 6	
3	Loud, about 3 sec., large black crickets in burrows 4	
3'	Soft sound, large populations of small crickets on the ground. . . 5	
4	May to July . . . . *Northern Spring Field Cricket*	
4'	July to first frost . . . . *Northern Fall Field Cricket*	
5	On ground, open fields, meadows, about 1.5 per sec.	
	. . . . *Striped Ground Cricket*	
5'	On ground, oak-hickory forests, about 5 per sec., tinkling	
	. . . . *Tinkling Ground Cricket*	
6	In high vegetation (e.g., goldenrod), bushes, trees 10	
6'	On the ground. 7	
7	In bogs, very faint trill . . . . *Sphagnum Ground Cricket*	
7'	Not in bogs. 8	
8	A stuttering, uneven trill, moist grassy area.	
	. . . . *Carolina Ground Cricket*	
8'	A more even trill. 9	
9	In dry sandy areas, a gray cricket. . . . . *Sand Ground Cricket*	

9'   In open fields, large populations............ *Allard's Ground Cricket*
10   Loud chorusing trills resembling spring peepers .... *Tree Crickets*
10' Pulsing, synchronous trills .... *Snowy Tree Cricket*
11   Crackling, rustling, shuffling sounds made in flight .... (locusts) 12
11' Sounds not made in flight... 15
12   Large locust fluttering in vertical flight .... *Black-Winged Locust*
12' Sounds made in horizontal flight............................. 13
13   Locusts in old fields, dry grassy areas, or bare areas........... 15
13' Locusts usually on rocky ledges 14
14   Exceptionally loud staccato crackle .... *Cracker Locust*
14' Rattling flight when chased, otherwise silent .... *Ledge Locust*
15   Yellow-wings, short, erratic, circular flights....................
     .... *Sulphur-Winged Locust*
15' Dull-colored body, straight flight................... *Clouded Locust*
16   Buzzing, lisping, continuous or interrupted.................. 17
16' Simple, infrequent squawks or buzzes ....
     .... (*See descriptions of katydid songs*)
17   Faint, dull, infrequent buzzes made by rubbing legs and wings,
     locusts on bare ground or in grass 18
17' Moderate long to continuous buzzing, lisping, wheezing...... 21
18   Extremely faint dull tattoo made by stamping on ground, also
     pulses made by rubbing legs and wings .... *Clouded Locust*
18' Soft tsikk, tsikk, made by rubbing legs and wings 19
19   Usually restricted to bunch grass.............. *Bunch-Grass Locust*
19' Not restricted to bunch grass 20
20   Series of soft buzzes, tsikk, tsikk, tsikk .... *Sprinkled Locust*
20' More lisping, soft series of syllables .... *Marsh Meadow Locust*
21   Continuous, as loud as a cicada, late afternoon and night.........
     .... *Robust Conehead*
21' Less loud, interrupted trains of buzzing or lisping 22
22   Extremely faint, inaudible beyond one or two feet............. 23
22' Loud enough to be heard for 10 feet or more................. 25
23   Moderately long trains of sound, no clicks, a soft hushing gentle
     song heard in salt marshes........ *Salt-Marsh Meadow Grasshopper*
23' Short trills of faint buzzing interrupted by clicks 24
24   One to five clicks between buzzes...............................
     .... *Short-Winged Meadow Grasshopper*

24′ Never five clicks between buzzes, incessant wheezing day and night, tse-tse-tse-pseee . . . . *Slender Meadow Grasshopper*

25 Rapid series of lisps tzip-tzip-tzip for minutes at a time. . . . . *Sword-Bearer Conehead*

25′ More buzzing than lisping 26

26 With clicks 27

26′ Clicks usually absent . . . . *Gladiator Meadow Grasshopper*

27 Buzzes each lasting 2 to 3 sec., followed by 6 to 9 clicks . . . . *Dusky-Faced Meadow Grasshopper*

27′ Similar to *Gladiator* but preceded and followed by clicks, day and night . . . . *Common Meadow Grasshopper*

Katydid and Shield-Bearer Songs

True Northern Katydid   Very loud, harsh, 2- to 3-pulse phrase, one per second, most often chorusing colonies in trees at night.

Texan Bush Katydid   A series of irregularly spaced pulses making a song without any regular pattern, a soft sh-sh-sh-sh. Rarely a zeet-zeet-zeet-zeet. Afternoon and evening. Low meadows and marshes.

Broad-Winged Bush Katydid   Day song an occasional zip or crick, followed by a long drawn-out exceedingly loud cr-r-r-rick several times in succession. Night call a repeated series of zikk-zikk-zikk-zikk.

Northern Bush Katydid   Trains of approximately ten soft long (2–3 sec.) pulses, with many clicks between trains.

Curve-Tailed Bush Katydid   Day song sequences of one to three pulses, bzriwi-bzriwi-bzriwi-bzriwi, more frequent than *Fork-Tailed* species. Night song tchw-tchw-tchw, about eight times. Trains about five seconds apart. Marshy meadows, low bushes.

Fork-Tailed Bush Katydid   A soft lisping zeep or zeep-zeep-zeep at irregular intervals, separated by clicks. Mostly late afternoon, less frequently at night. Bushes and tall grasses bordering woods.

Round-Winged Katydid   Fast, soft, lisping buzz tsip-i-tsip-i-tsip-i, continuing indefinitely. Also groups of rattly buzzy phrases, chick-a-chee——chick-a-chee. Day and night. Old fields, grass, weeds, shrubs, woodland borders.

Oblong-Winged Katydid   A harsh, scraping kizizik (chirp) at irregular intervals. A phrase sounding like z-z-z-z-zik-zik. At night. Shrubbery, goldenrod.

Short-Legged Shield Bearer   A group of soft buzzes each lasting 2–5 seconds or longer. A brief pause between each group. Has a lisping character (sh-sh-sh-sh——sh-sh-sh) similar to the song of the *Common Meadow Grasshopper*. Mostly at night. In leaf litter in open woods.

Long-Legged Shield Bearer   A continuous train of soft impulses. At night. In leaf litter in open woods.

References

Alexander, R. D. 1957. Sound production and associated behavior in insects. *Ohio Journal of Science* 57: 101–113.

———— 1960. Sound communication in Orthoptera and Cicadidae. In *Animal Sounds and Communication*, vol. 7, pp. 38–92. American Institute of Biological Science Publications.

———— 1961. Aggressiveness, territoriality, and sexual behavior in field crickets (Orthoptera: Gryllidae). *Behaviour* 17: 130–223.

———— 1967. Seasonal and daily chirping cycles in the northern spring and fall field crickets, *Gryllus veletis* and *G. pennsylvanicus*. *Ohio Journal of Science* 67: 200–209.

————, A. E. Pace, and D. Otte. 1972. The singing insects of Michigan. *Great Lakes Entomologist* 5: 33–69.

Alexander, R. D., and E. S. Thomas. 1959. Systematic and behavioral studies on the crickets of the Nemobius fasciatus group (Orthoptera: Gryllidae, Nemobiinae). *Annals of the Entomological Society of America* 52: 591–605.

Blatchley, W. S. 1920. *Orthoptera of Northeastern America*. Indianapolis: Nature Publishing Co.

Cantrall, I. J. 1943. The ecology of the Orthoptera and Dermaptera of the George Reserve, Michigan. *University of Michigan Museum of Zoology Miscellaneous Publications* 54: 1–182.

Fernald, C. H. 1888. *The Orthoptera of New England*.

Fulton, B. B. 1915. The crickets of New York: Life history and bionomics. *New York Agricultural Experiment Station Bulletin* 42: 1–47.

———— 1931. A study of the genus Nemobius (Orthoptera: Gryllidae). *Annals of the Entomological Society of America* 24: 205–237.

———— 1951. The seasonal succession of orthopteran stridulation near Raleigh, North Carolina. *J. Elisha Mitchell Scientific Society* 67: 87–95.

Morris, G. K., and T. V. Walker. 1976. Calling songs of *Orchelimum* meadow katydids (Tettigoniidae). I: Mechanism, terminology, and geographic distribution. *Canadian Entomological Society* 108: 785–800.

Morse, A. P. 1920. Manual of the Orthoptera of New England. *Proceedings of the Boston Society of Natural History* 35: 197–556.

———— 1921. Orthoptera of Maine. *Maine Agricultural Experiment Station Bulletin* 296: 1–35.

Otte, D. 1970. A comparative study of communicative behavior in grasshoppers. *University of Michigan Museum of Zoology Miscellaneous Publications* 141: 1–168.

———— 1981. *The North American Grasshopper.* Vol. I. Cambridge, Mass.: Harvard University Press.

———— 1984. *The North American Grasshopper.* Vol. II. Cambridge, Mass.: Harvard University Press.

Pierce, G. W. 1948. *The Songs of Insects, with Related Material on the Production, Propagation, Detection, and Measurement of Sonic and Supersonic Vibrations.* Cambridge, Mass.: Harvard University Press.

Piers, H. 1918. The Orthoptera (cockroaches, locusts, grasshoppers, and crickets) of Nova Scotia, with descriptions of the species and notes on their occurrence and habits. *Transactions of the Nova Scotia Institute of Science* 14: 201–356.

Scudder, S. H. 1892. The songs of our grasshoppers and crickets. *Annual Report of the Entomological Society of Ontario, 1891* (22): 62–78.

Thomas, E. S., and R. D. Alexander. 1962. Systematic and behavioral studies on the meadow grasshoppers of the *Orchelimum concinnum* group (Orthoptera: Tettigoniidae). *University of Michigan Museum of Zoology Occasional Papers* 626: 1–31.